WASTEWATER COLLECTION SYSTEMS MANAGEMENT

Prepared by the **Wastewater Collection Systems Management Task Force** of the **Water Environment Federation**

S. Wayne Miles, P.E., BCEE, *Chair*

Abraham Araya, Ph.D.	Roger W. Lehman, P.E.
Nick Arhontes, P.E., BSCE	Robert Lockridge
Hal Balthrop, P.E.	John M. Mastracchio, P.E., CFA
Samantha E. Bartow	Carol Malesky
Robert Beringer, P.E.	Robert L. Matthews, P.E. BCEE
Andy Bowen, P.E.	Susan R. McHugh
James N. Broome, P.E.	Amy Taylor-McMillian
Gregory R. Chol, P.E.	Jack Moyer
Bruce J. Corwin, P.E.	Sean P. Murphy, P.E.
Rich Cunningham	Aaron K. Nelson, P.E.
Ted DeBoda, P.E.	Paul Pinault, P.E.
Robert Decker, P.E.	Alison Ratliff
James M. Faccone	Ross E. Schlobohm, P.E.
Rudy Fernandez, P.E.	Kelly D. Shephard
Brent Freeman, P.E.	Marsha W. Slaughter, P.E.
William W. S. Gray	Peter L. Stump, P.E.
David Guhin, P.E.	Richard Thomasson
Rob Horvat	Philip Tiewater, P.E.
Richard E. Huffman, P.E.	Stephen Tilson
James Joyce, P.E.	Larry G. Tolby
Jennifer Kauffman	Marc P. Walch, P.E.
	Tina Wolff

Under the Direction of the **Collection Systems Subcommittee** of the **Technical Practice Committee**

2009

Water Environment Federation
601 Wythe Street
Alexandria, VA 22314-1994 USA
http://www.wef.org

WASTEWATER COLLECTION SYSTEMS MANAGEMENT

WEF Manual of Practice No. 7
Sixth Edition

*Prepared by the Wastewater Collection Systems Management Task Force
of the Water Environment Federation*

WEF Press

Water Environment Federation   Alexandria, Virginia

New York   Chicago   San Francisco   Lisbon   London   Madrid
Mexico City   Milan   New Delhi   San Juan   Seoul
Singapore   Sydney   Toronto

The McGraw·Hill Companies

Cataloging-in-Publication Data is on file with the Library of Congress.

McGraw-Hill books are available at special quantity discounts to use as premiums and sales promotions, or for use in corporate training programs. To contact a representative please e-mail us at bulksales@mcgraw-hill.com.

Wastewater Collection Systems Management: WEF Manual of Practice No. 7, Sixth Edition

Copyright © 2010 by the Water Environment Federation. All rights reserved. Printed in the United States of America. Except as permitted under the United States Copyright Act of 1976, no part of this publication may be reproduced or distributed in any form or by any means, or stored in a data base or retrieval system, without the prior written permission of the publisher.

1 2 3 4 5 6 7 8 9 0   FGR/FGR   0 1 4 3 2 1 0 9

ISBN 978-0-07-166663-3
MHID 0-07-166663-X

The sponsoring editor for this book was Larry S. Hager and the production supervisor was Pamela A. Pelton. It was set in Palatino by Lone Wolf Enterprises, Ltd. The art director for the cover was Jeff Weeks.

Water Environment Research, WEF, and *WEFTEC* are registered trademarks of the Water Environment Federation.

This book is printed on acid-free paper.

IMPORTANT NOTICE

The material presented in this publication has been prepared in accordance with generally recognized engineering principles and practices and is for general information only. This information should not be used without first securing competent advice with respect to its suitability for any general or specific application.

The contents of this publication are not intended to be a standard of the Water Environment Federation (WEF) and are not intended for use as a reference in purchase specifications, contracts, regulations, statutes, or any other legal document.

No reference made in this publication to any specific method, product, process, or service constitutes or implies an endorsement, recommendation, or warranty thereof by WEF.

WEF makes no representation or warranty of any kind, whether expressed or implied, concerning the accuracy, product, or process discussed in this publication and assumes no liability.

Anyone using this information assumes all liability arising from such use, including but not limited to infringement of any patent or patents.

About WEF

Formed in 1928, the Water Environment Federation (WEF) is a not-for-profit technical and educational organization with 35,000 individual members and 81 affiliated Member Associations representing an additional 50,000 water quality professionals throughout the world. WEF and its member associations proudly work to achieve our mission of preserving and enhancing the global water environment.

For information on membership, publications, and conferences, contact

Water Environment Federation
601 Wythe Street
Alexandria, VA 22314-1994 USA
(703) 684-2400
http://www.wef.org

Contents

PREFACE xix
LIST OF FIGURES xxiii
LIST OF TABLES xxv

Chapter 1 Introduction 1
1.0 PURPOSE 1
2.0 TARGET AUDIENCE 2
3.0 ASSET MANAGEMENT AND CAPACITY, MANAGEMENT,
    OPERATIONS, AND MAINTENANCE GUIDING PRINCIPLES 3
   3.1 Asset-Management Principles 3
   3.2 Capacity, Management, Operations, and Maintenance
       Program Principles 9
4.0 REFERENCES 10
5.0 SUGGESTED READINGS 11

Chapter 2 System Operations and Maintenance 13
1.0 INTRODUCTION 14
2.0 PROGRAM OBJECTIVES 14
3.0 ESTABLISHING OR EVALUATING AN OPERATIONS
    AND MAINTENANCE PROGRAM 15
   3.1 System Asset Characteristics 15
   3.2 Performance History 16
   3.3 Agency Resources 16
   3.4 Customer Expectations/Complaints 17

3.5 Community Interests 18
3.6 Regulatory Resources 19
4.0 PROGRAM COMPONENTS 19
    4.1 Maintenance 23
    4.2 Mapping Use 27
    4.3 Directed Maintenance Programs 27
    4.4 Application of Technology to Support Operations
        and Maintenance 29
    4.5 System Problems and Overflow Response 31
5.0 PUMPING STATION OPERATIONS
    AND MAINTENANCE PROGRAMS 34
    5.1 Increasing Importance of Pumping Station Maintenance 36
    5.2 Types of Pumping Station Maintenance 37
6.0 PUMPING STATION OPERATIONS
    AND MAINTENANCE STAFFING 40
    6.1 Pumping Station Operations Staffing 42
    6.2 Pumping Station Maintenance Staffing 43
7.0 OPERATIONS AND MAINTENANCE FACILITIES
    AND EQUIPMENT 44
8.0 REFERENCES 46

Chapter 3 Information Management 47

1.0 INTRODUCTION 48
2.0 ESTABLISHING AN INFORMATION MANAGEMENT SYSTEM 48
    2.1 Conformance to Regulatory Responsibilities 49
    2.2 Performing Timely and Consistent Information Searches 50
    2.3 Maintaining Perpetual Calendars for Preventive
        Maintenance and Inspections 51
    2.4 Providing Justification for Operations Budgets 51
    2.5 Tracking Prioritized Work Order and Repair Schedules 52
    2.6 Organizing Capital Rehabilitation and Replacement
        Plans Based on Asset-Management Data 52
    2.7 Preserving the Utility's Collection System Corporate Memory ... 53

3.0 BUILDING AN INFORMATION MANAGEMENT SYSTEM 53
  3.1 Types of Collection System Information 53
  3.2 Selecting the System Attributes to Retain in a Collection System Information Management System 54
  3.3 Practical Guidelines for Establishing an Information Management System 57
4.0 OPERATING AND MAINTAINING AN INFORMATION MANAGEMENT SYSTEM 58
5.0 REFERENCE 58

Chapter 4 Collection System Assessment and Capital Improvement Planning 59

1.0 INTRODUCTION 60
2.0 SYSTEM ANALYSIS 61
  2.1 Capacity Assurance Planning 61
  2.2 System Analysis Tools and Methods 62
    2.2.1 Data Collection and Management Methods 62
    2.2.2 Supervisory Control and Data Acquisition Systems 63
    2.2.3 Geographic Information Systems and Computer-Assisted Drawing 63
    2.2.4 Computerized Maintenance Management Systems 64
      2.2.4.1 Maintenance History 64
      2.2.4.2 Complaints and Service Requests 65
      2.2.4.3 Work Orders 66
    2.2.5 Integrating Geographic Information Systems and Computerized Maintenance Management Systems 66
    2.2.6 Inflow and Infiltration Identification 67
    2.2.7 Flow Monitoring/Monitoring 69
    2.2.8 Hydraulic and Hydrologic Modeling 72
      2.2.8.1 Steady-State Simulation versus Dynamic Simulation .. 73
      2.2.8.2 Flow Development 74
      2.2.8.3 Calibration/Verification 75
  2.3 Condition Assessment 76

        2.3.1 Surface Inspection 77
        2.3.2 Closed-Circuit Television 78
        2.3.3 Smoke Testing 79
        2.3.4 Trunkline Inspection 80
        2.3.5 Lift Stations 81
3.0 CAPITAL PLANNING 82
    3.1 Master Planning 84
    3.2 Capital Improvement Plan Development 85
        3.2.1 Ratings. 85
            3.2.1.1 Condition 88
            3.2.1.2 Performance 88
            3.2.1.3 Risk 88
        3.2.2 Validation. 90
        3.2.3 Replacement/Rehabilitation Goals 91
        3.2.4 Industry Benchmarks. 91
        3.2.5 Prioritization 92
        3.2.6 Final Product 93
4.0 REFERENCES 94
5.0 SUGGESTED READINGS 94

Chapter 5 System Design Considerations 97

1.0 INTRODUCTION 99
    1.1 Regulatory and Environmental Requirements 100
        1.1.1 National Pollutant Discharge Elimination System Permits 100
            1.1.1.1 Sanitary Sewer Overflows and Combined Sewer Overflows 100
            1.1.1.2 Capacity Management, Operations, and Maintenance 100
        1.1.2 System Conveyance Design Rates. 101
            1.1.2.1 Risk-Based Design Options 102
            1.1.2.2 Wet-Weather Facilities 102
        1.1.3 Design Approvals 102
        1.1.4 Environmental Issues and Compliance. 103

1.2 Permitting	103
2.0 DESIGN GUIDELINES	103
2.1 Design Process Checklist	103
2.2 Project Management	104
2.3 Using Consultants	105
2.4 Routing Considerations	105
2.4.1 Assessing Aboveground and Belowground Interference	*105*
2.4.1.1 Surveys	*106*
2.4.1.2 Potholing	*106*
2.4.2 Geotechnical	*106*
2.4.3 Traffic Control	*106*
2.4.4 Permanent Access	*107*
2.4.5 Public Effects	*107*
2.4.6 Rights-of-Way and Easements	*108*
2.5 Hydraulic Analysis Using Modeling	108
2.6 Pumping Stations	109
2.6.1 Package Stations	*109*
2.6.2 Other Types of Stations	*109*
2.7 Pipelines and Joint Materials	110
2.7.1 Gravity Pipelines	*110*
2.7.1.1 Sizing Considerations	*110*
2.7.1.2 Materials	*111*
2.7.1.3 Backfill and Bedding	*112*
2.7.1.4 Grades	*114*
2.7.1.5 Inflow and Infiltration Allowance	*114*
2.7.2 Pressure Pipelines	*115*
2.7.2.1 Types of Pressure Systems	*115*
2.7.2.2 Sizing Considerations	*116*
2.7.2.3 Materials	*116*
2.7.2.4 Backfill	*117*
2.7.2.5 Restrained Joints	*117*

   2.7.2.6 *Blowoff and Air Valves* 118
  2.7.3 *Odor and Corrosion Control* 119
   2.7.3.1 *Sulfide Control Design Considerations* 120
   2.7.3.2 *Corrosion Protection Design Considerations* 122
 2.8 Rehabilitation 123
  2.8.1 *Condition Assessment for Design* 123
  2.8.2 *Sewer System Assessment Protocols* 123
  2.8.3 *Structural Repairs* 124
  2.8.4 *Technologies for Design* 124
  2.8.5 *Inflow and Infiltration* 125
 2.9 Manholes 125
  2.9.1 *Spacing* 126
  2.9.2 *Materials* 127
   2.9.2.1 *New Construction* 127
   2.9.2.2 *Rehabilitation* 127
  2.9.3 *Channels* 128
  2.9.4 *Covers* 128
  2.9.5 *Steps* 129
  2.9.6 *Drop Manholes* 129
  2.9.7 *Connection between Manhole and Sanitary Sewer* 130
  2.10 Manhole and Pipeline Testing 130
   2.10.1 *New Construction* 130
   2.10.2 *Rehabilitation* 131
  2.11 Inverted Siphons 132
  2.12 Sanitary Sewers Above Grade 133
  2.13 Crossings and Tunnels 133
   2.13.1 *Stream Crossings* 133
   2.13.2 *Other Crossings* 133
  2.14 Service Connections and Disconnections 134
  2.15 Value Engineering 135
  2.16 Specification Writing 135
3.0 REFERENCES 136

Chapter 6 Construction Contracting 137

1.0 CONSTRUCTION PROJECTS AND PROJECT MANAGEMENT 138
2.0 LAWS AND REGULATIONS 139
    2.1 Contracting Requirements 139
    2.2 Competitive Bidding 140
    2.3 Environmental Requirements 141
    2.4 Interagency Permit Requirements 141
3.0 CAPITAL PROJECT DELIVERY METHODS 142
    3.1 Design–Bid–Build 142
    3.2 Multiple Prime Contractors 142
    3.3 Design–Build 143
    3.4 At-Risk Construction Management 144
4.0 AGENCY CONSTRUCTION MANAGEMENT 145
5.0 MAINTENANCE PROJECT DELIVERY METHODS 145
    5.1 Force Account Construction 145
    5.2 Term Contracts 145
6.0 CONSTRUCTION CONTRACT DOCUMENTS 146
    6.1 Constructability Review 146
    6.2 Operations and Maintenance Review 146
    6.3 Value Engineering 146
    6.4 Allocation of Risk 147
        6.4.1 Unforeseen Conditions............................ 147
        6.4.2 Dispute Resolution 148
        6.4.3 Conflict Resolution 148
        6.4.4 Escrow Documents 148
7.0 FINANCIAL CONSIDERATIONS 148
    7.1 Payment Methods for Construction Contracts 148
        7.1.1 Lump-Sum Basis 149
        7.1.2 Unit-Price Basis................................. 149
        7.1.3 Cost-Plus Basis 149
    7.2 Funding Mechanisms 150

  7.2.1 *Municipal Bonds (Local Financing)* *150*
  7.2.2 *Federal and State Financing (Grants and Loans)* *150*
 7.3 Cash Flow/Cost Control 151
 7.4 Construction Bonds 151
 7.5 Construction Insurance 151
8.0 QUALITY ASSURANCE/QUALITY CONTROL 152
 8.1 Construction Inspection/Administration 153
  8.1.1 *Field Directives* *153*
  8.1.2 *Engineering Bulletins* *153*
 8.2 Schedule 153
 8.3 Post-Construction Closed-Circuit Television Inspection 153
 8.4 Warranty/Guarantee Administration 154
9.0 SAFETY CONTROL METHODS AND PROCEDURES 154
10.0 UTILITY DAMAGE CONTROL (ONE-CALL SYSTEMS) 154
11.0 ALLIANCES AND PARTNERING 155
12.0 REFERENCES 155
13.0 SUGGESTED READINGS 157

Chapter 7 Public Policy and Community Relations 159

1.0 PUBLIC POLICY THEORY 160
2.0 PUBLIC POLICY IN WASTEWATER COLLECTION
  SYSTEM MANAGEMENT 161
3.0 ESTABLISHING WASTEWATER COLLECTION PUBLIC POLICY .. 162
4.0 KEEPING PUBLIC POLICY RELEVANT 163
5.0 PUBLIC POLICY AND COMMUNITY RELATIONS 165
6.0 COMMUNITY RELATIONS AND PUBLIC EDUCATION 166
7.0 COMMUNITY RELATIONS AND PUBLIC MEETINGS 167
8.0 COMMUNITY RELATIONS AND CUSTOMER SERVICE 169
9.0 COMMUNITY RELATIONS AND EMERGENCY PREPAREDNESS . 171
10.0 REFERENCES 171
11.0 SUGGESTED READINGS 171

Chapter 8 Budgeting and Financial Planning............ 173
1.0 INTRODUCTION 174
2.0 OPERATIONS BUDGET 175
3.0 CAPITAL IMPROVEMENT BUDGET 179
4.0 SETTING BUDGET PRIORITIES THROUGH ASSET
    MANAGEMENT 181
    4.1 The Problem 182
    4.2 The Solution 182
    4.3 Underground Infrastructure: The Tough Asset ... 183
    4.4 Life-Cycle Cost Analysis 185
        4.4.1 Equivalent Annual Cost Approach 186
        4.4.2 Common-Life Approach 186
        4.4.3 Steps to Completing a Life-Cycle Cost Analysis ... 187
5.0 CAPITAL FUNDING AND FINANCING OPTIONS 188
    5.1 Capital Funding Sources 188
        5.1.1 User Charges 188
        5.1.2 Ad Valorem Taxes 189
        5.1.3 Special Assessments 189
        5.1.4 Impact Fees 189
        5.1.5 Grants 189
    5.2 Capital Financing Sources 190
        5.2.1 General Obligation Bonds 190
        5.2.2 Revenue Bonds 190
        5.2.3 Governmental Loans 191
6.0 REVENUE REQUIREMENTS AND FEE SETTING 191
    6.1 Financial Condition 192
    6.2 Establishing the Basis for Fees and Charges .. 193
        6.2.1 Cash-Needs Approach 193
        6.2.2 Utility Approach 194
        6.2.3 Selecting a Test Year 195
        6.2.4 User Charge Revenue Requirements 195

  6.3 User Charge and Service Fee Types 196
    6.3.1 User Charges... 196
    6.3.2 Impact Fees... 197
    6.3.3 Other Fees 198
7.0 ROLE OF GOVERNMENT ACCOUNTING STANDARDS BOARD STATEMENT NUMBER 34 IN BUDGETING AND FINANCIAL PLANNING 198
8.0 SELLING THE ASSET MANAGEMENT APPROACH AND BUDGET PLAN 200
9.0 REFERENCES 202

Chapter 9 Safety, Standard Procedures, Training, and Certifications 203

1.0 INTRODUCTION TO AND APPLICABILITY OF THE OCCUPATIONAL HEALTH AND SAFETY ADMINISTRATION ... 203
2.0 APPLICABILITY OF HEALTH AND SAFETY REGULATIONS 204
3.0 HEALTH AND SAFETY PROGRAM ELEMENTS 204
  3.1 Health and Safety Policy 205
  3.2 Individual Responsibility 207
  3.3 Safety Committee 207
  3.4 Health and Safety Rules and Written Programs 208
  3.5 Standard Operating Procedures 210
  3.6 Employee Orientation and Training 210
  3.7 Workplace Inspections 212
  3.8 Emergency Medical and First-Aid Procedures 213
  3.9 Medical Aid and First Aid 213
  3.10 Health and Safety Program Promotion 214
4.0 SAFETY PROGRAM ORGANIZATION 214
5.0 SAFETY EQUIPMENT 215
6.0 SUMMARY 216
7.0 REFERENCES 216

Chapter 10 Emergency Preparedness and Security 217

1.0 INTRODUCTION AND OVERVIEW 218
2.0 DISASTER AND SECURITY RISKS AND THEIR POTENTIAL EFFECTS ON WASTEWATER COLLECTION SYSTEMS 219
    2.1 Natural Disasters 219
    2.2 Humanmade Incidents 219
    2.3 Technological Failures 220
3.0 PREPAREDNESS PLANNING ACTIVITIES FOR WASTEWATER COLLECTION SYSTEMS 220
    3.1 Develop an Organizational Culture of Emergency Preparedness and Security 221
    3.2 Prepare Employees for Disaster Incidents 221
    3.3 Develop Viable Emergency Response Plans 222
    3.4 Establish Relationships with Other Local and State Response Agencies 223
    3.5 Develop an Understanding of the National Incident Management System and the Incident Command System 223
    3.6 Be Aware of Cyber-Vulnerabilities 224
    3.7 Enter into Mutual Aid Networks and Agreements 224
    3.8 Develop Crisis Communications Plans 225
    3.9 Conduct Training and Exercises 225
    3.10 Stay Abreast of Developments in the Industry 226
4.0 PHYSICAL PREPAREDNESS FOR WASTEWATER COLLECTION SYSTEMS 226
    4.1 Provide for Emergency Electrical Power 227
    4.2 Ensure Adequate Protection of Mains from Waterways 227
    4.3 Reinforce Vulnerable Mains 228
    4.4 Locate Pumping and Lift Stations above Flood Elevations ... 228
    4.5 Protect Vehicles and Heavy Equipment 229
    4.6 Provide for Necessary Repair Equipment and Materials 229
    4.7 Ensure the Availability of Reliable Communication Systems ... 229

5.0 DISASTER RESPONSE IN WASTEWATER
COLLECTION SYSTEMS 230
  5.1 Activate Emergency Response Plans and Open Emergency
      Operations Centers 230
  5.2 Activate Crisis Communications Plans 230
  5.3 Check on the Welfare of Employees 231
  5.4 Conduct Damage Assessments 231
  5.5 Communicate and Coordinate with Other Local and
      State Response Agencies 231
  5.6 Activate Mutual Aid Networks and Agreements 232
  5.7 Make Emergency Repairs 232
  5.8 Maintain Records to Maximize Federal Emergency
      Management Agency Reimbursement 232
6.0 REFERENCES 233

INDEX 235

Preface

For many local governments, the wastewater collection system is one of the most valuable assets they own. In addition to being a valuable physical asset, the value wastewater collection systems provide to protect public health and the environment is tremendous. Too often, however, this value is under-recognized and under-appreciated by the community. This can lead to inadequate reinvestment in the system and insufficient resources to provide for proper management, operations, and maintenance of the system.

To address these challenges, wastewater collection system managers must be exceptionally proficient in a multitude of technical and non-technical skills needed to efficiently and effectively operate and maintain a collection system. *Wastewater Collection Systems Management* (MOP 7; sixth edition) provides guidance, strategies, and ideas that may be used by a system manager to be successful within these many fields of expertise.

Since the 1999 version of this manual, the regulatory focus on wastewater collection systems has increased across the United States. This is evidenced by the issuance of numerous enforcement actions by the U.S. Environmental Protection Agency (U.S. EPA) and state water- quality agencies against wastewater utilities, mandating increased system investment and program improvements. In addition, U.S. EPA has developed capacity, management, operations, and maintenance program guidelines that outline best practices for wastewater collection system managers to use to improve overall system performance. These guidelines are referenced in numerous chapters of this manual and offer a good basis for assessing whether a wastewater collection system program is comprehensive and sustainable.

Asset-management principles have become more widely applied to wastewater collection systems than was true in 1999. Though the 1999 version of this manual included a chapter titled Capital Asset Management, this version has integrated asset-management principles to individual topics throughout the manual. For example, a new chapter titled Information Management has been added to this manual to emphasize the importance that having access to good data has to collection system management and to provide guidance on how managers can leverage digital tools to improve decision-making. The chapter on Collection System Assessment and Capital Improvement Planning has been added to reflect the importance of understanding system condition and incorporating this knowledge and other aspects, such as system capacity needs, to the development of a capital improvements plan. Additionally, the budget and financial aspects of asset management have been divided into a new, separate chapter so that these important considerations can be more readily referenced.

In light of a number of natural and humanmade disasters that have occurred in the United States in recent years—including Hurricane Katrina hitting the Gulf Coast in 2005 and the terrorist attacks of September 11, 2001, in the United States—increased attention has been given to the need for utilities managers to prepare and plan for similar events. As a result, this manual includes a chapter titled Emergency Preparedness and Security that is new to this edition.

The sixth edition of this Manual of Practice was produced under the direction of S. Wayne Miles, P.E., BCEE, *Chair*. The principal authors of this Manual of Practice are as follows:

Chapter 1	S. Wayne Miles, P.E., BCEE
Chapter 2	Hal Balthrop, P.E.
Chapter 3	Richard Cunningham
Chapter 4	David Guhin, P.E.
Chapter 5	Bruce J. Corwin, P.E.
Chapter 6	Sean P. Murphy, P.E.
Chapter 7	Tina Wolff
Chapter 8	John M. Mastracchio, P.E., CFA
Chapter 9	Amy Taylor-McMillian
Chapter 10	Jack Moyer

Contributing authors and the chapters to which they contributed are Robert Lockridge (8) and Carol Malesky (8).

In addition to the WEF Task Force and Technical Practice Committee Control Group members, reviewers include Cil Pierce, Marcia S. Maurer, Jacob Boomhouwer, and Chris Sorensen.

Authors' and reviewers' efforts were supported by the following organizations:

Aurora Water Capital Projects Division, Aurora, Colorado
Bergmann Associates, PC, Rochester, New York

CDM, Rancho Cucamonga, California; Rancho Murieta, California; Fort Myers, Florida; Raleigh, North Carolina; and Portland, Oregon
City of Edmonton Drainage Services, Alberta, Canada
City of Great Falls Public Works Department, Montana
City of Greensboro Water Resource, North Carolina
City of Portland, Bureau of Environmental Services, Oregon
City of Santa Rosa Utilities Department, California
CTE, Kansas City, Missouri
E L M Consulting, Wayne, Pennsylvania
Granite Quarry, North Carolina
Hazen and Sawyer, P.C., Chesapeake, Virginia
HDR Engineering, Las Vegas, Nevada, and Bellevue, Washington
King County/DNRP Wastewater Treatment Division, Seattle, Washington
KLH Engineers, Inc., Pittsburgh, Pennsylvania
Malcolm Pirnie, Inc., Maitland, Florida; Latham, New York; Akron, Ohio; and Cleveland, Ohio
Nashville Metro Water Services, Nashville, Tennessee
National Clay Pipe Institute, Modesto, California
O'Brien & Gere, Rochester, New York
Odor and Corrosion Technology Consultants, Inc., Houston, Texas
Orange County Sanitation District, Fountain Valley, California
Orange County Utilities, Orlando, Florida
Parsons Brinckerhoff, West Palm Beach, Florida; Atlanta, Georgia; and Herndon, Virginia
PBS&J, Orlando, Florida
Sacramento Regional County Sanitation District, California
Tilson & Associates LLC, Torrington, Connecticut
Township of Hampton, Allison Park, Pennsylvania
URS Corporation, Wilmington, Delaware; Baltimore, Maryland; Charlotte, North Carolina; and Morrisville, North Carolina
Village of Ruidoso, New Mexico
Western Carolina Regional Sewer Authority, Greenville, South Carolina
Weston Solutions, Inc., West Chester, Pennsylvania

List of Figures

Figure		Page
4.1	Typical combined sewer system schematic	71
4.2	Capital improvement plan prioritization overall scoring methodology	86
4.3	Capital improvement plan prioritization for sanitary sewer pipe condition	87
4.4	Typical capital improvement plan layout and components	89
7.1	The five steps of the public policy cycle	160
8.1	Rate effects of different asset-management approaches	184
8.2	The financial management cycle	192
9.1	Sample standard operating procedure format	206
9.2	Sample job hazard analysis form	208

List of Tables

Table   **Page**

1.1   Glossary of terms 6

8.1   Example statement of the effect on the current service level of an increase or decrease in expenditures as represented by the baseline budget 178

WASTEWATER COLLECTION SYSTEMS MANAGEMENT

Chapter 1

Introduction

1.0	PURPOSE	1	3.1	Asset-Management Principles	3
2.0	TARGET AUDIENCE	2	3.2	Capacity, Management, Operations, and Maintenance Program Principles	9
3.0	ASSET MANAGEMENT AND CAPACITY, MANAGEMENT, OPERATIONS, AND MAINTENANCE GUIDING PRINCIPLES	3	4.0	REFERENCES	10
			5.0	SUGGESTED READINGS	11

1.0 PURPOSE

The purpose of this Manual of Practice (MOP 7, 6th Edition) is to provide guidance and direction to those individuals responsible for the management and oversight of wastewater collection systems. In most urban and suburban communities, the wastewater collection system—which includes gravity sewers, pumping stations, force mains, and other sewer conveyance related facilities—is one of the largest, most valuable infrastructure assets. This system plays a vital role in the economic, social, and environmental viability of the community. It is important that the community recognizes the value of this system so that sufficient resources are dedicated to management, operations, and maintenance.

Managing a wastewater collection system poses many challenges. Aging and deteriorating infrastructure decreases system capacity and system reliability and increases operations and maintenance costs and risk of failure. Changes in wastewater flows, either from growth or a loss of significant customers, can make capacity needs difficult to manage. Third-party damage to the system and customer wastewater flows (both legal and illicit) can affect the system in ways that are difficult to predict.

A wastewater collection system manager faces many technical, financial, social, legal, and environmental challenges. Increased costs for energy, materials, and labor are stretching capital and operations budgets. A lack of understanding by most customers of how a wastewater collection system works results in confusion related to how revenues are spent and how this relates to service expectations. Managers also must plan for intensified regulatory scrutiny and possible risk of third-party litigation.

To be successful, a wastewater collection system manager requires knowledge of business, engineering, operations, maintenance, human resource management, and communications. This manual is intended to act as a reference guide on these many, varied aspects of managing collection systems; though it is not a complete body of knowledge on any of these topics. Additional references and lists of further readings are provided at the end of each chapter.

2.0 TARGET AUDIENCE

Although this MOP is directed primarily at wastewater collection system managers and superintendents, many others will benefit from it. Utility engineers, system operators, local elected officials, municipal administrators, financial managers, municipal communications professionals, regulatory officials, consultants, or anyone with an interest in protecting public health and the environment will find it useful. Increasingly, the mainstream media is covering water and wastewater issues, which is helping to improve the public's understanding of the importance of infrastructure. Utility managers can and should leverage these opportunities to engage stakeholders in conversations about the importance of good wastewater system management and the value that it provides to the community.

Many cities are investing heavily in urban redevelopment projects aimed at reigniting the economic and social viability of their downtown business districts. These renewal projects offer an opportunity to engage local officials and business leaders in a discussion about the importance of rehabilitating the wastewater collection system to meet urban redevelopment and sustainability goals. Wastewater collection system managers and personnel must be active in the planning and implementation of these renewal projects to ensure long-term success because system rehabilitation and replacement will be vital to meeting performance objectives cost effectively.

3.0 ASSET MANAGEMENT AND CAPACITY, MANAGEMENT, OPERATIONS, AND MAINTENANCE GUIDING PRINCIPLES

The U.S. Environmental Protection Agency (U.S. EPA) has been a leading advocate in the application of asset-management principles in managing our country's wastewater infrastructure (Albee, 2005). Applying asset-management principles requires utility managers to make decisions based on life cycle cost analyses, risk assessments using condition and criticality factors, and infrastructure asset valuation effects, among others. As of the publication of this manual, no national rules have been formally promulgated requiring the implementation of capacity, management, operations, and maintenance (CMOM) programs. Many utilities, however, are adopting these programs either because of consent-decree requirements or because of the improved system performance that can be achieved by their implementation. The principles of asset management and those promoted by CMOM programs are consistent with one another, as described further in the following sections. Although CMOM has not been formally promulgated, it is important to note that the wastewater collection system is a significant component of a wastewater treatment plant as defined the federal Clean Water Act, and thus is regulated at the federal level through the issuance of National Pollutant Discharge Elimination System (NPDES) permits.

3.1 Asset-Management Principles

Because a wastewater collection system is a group of critical infrastructure assets, and because this manual is focused on wastewater collection system management, in many ways the entire publication is about asset management for the specific group of assets contained in a wastewater collection system. This perspective helps to demonstrate the many areas of expertise that must be considered in the development of an asset-management program.

Managers responsible for the planning, design, construction, operation, or financing of a wastewater collection system are already managing assets. But managing assets is not the same as having an asset-management program. Managing assets includes the complex decision-making process that the collection system manager must use every day to set priorities, consider alternatives, and make the most efficient use of available resources to accomplish program objectives. An asset-management program is a set of procedures and protocols that institutionalize, document, and preserve that decision-making process. With staff members retiring at an ever-increasing rate from utility workforces, the importance of documenting and preserving system knowledge and experience is crucial. Although many wastewater collection systems have similarities, site-specific differences between systems must be clearly documented and understood. An asset-management program has several benefits:

- Makes the decision-making process understandable and transparent to others;
- Provides consistent criteria for making those decisions and balancing competing needs and interests;
- Minimizes the long-term costs of system operations and maintenance;
- Defines acceptable levels of service to the customers;
- Creates data and information processes that improve future decision-making; and
- Establishes roles, goals, and metrics that can focus and motivate the entire organization toward more cost-effective operation.

An effective asset-management program will help determine the funding needed to meet the service expectations of stakeholders, which comprise customers, elected officials, regulatory officials, the environmental community, and any other entities having a stake in the performance of the wastewater collection system. Although identifying and documenting service expectations can be difficult, not understanding those expectations can lead to longer-term difficulties. If customer level of service expectations are not met because of frequent system failures or lack of adequate performance, customer complaints may lead to undesirable political attention or, in some cases, regulatory enforcement actions. By exceeding customer expectations, the utility may increase the cost of service to greater than needed, also leading to complaints. The key is to improve customers' understanding that the level of service is directly proportional to cost. Whether the wastewater collection system program is funded through rates, taxes, connection fees, grants, or any combination thereof, an asset-management program will provide local decision-makers with sound, understandable data illustrating the revenue that is needed to provide a given level of service and the consequences of not getting this revenue on a timely basis.

Of the many definitions of asset management, the wastewater industry has adopted the following: "An integrated set of processes to minimize the life-cycle costs of infrastructure assets, at an acceptable level of risk, while continuously delivering established levels of service" (NACWA et al., 2007). The Federal Highway Administration and the American Association of State Highway and Transportation Officials (1996) provide another relevant definition: "A systematic process of maintaining, upgrading, and operating physical assets cost effectively. It combines engineering principles with sound business practices and economic theory, and it provides tools to facilitate a more organized, logical approach to decision making. Thus, asset management provides a framework for handling both short and long-range planning."

Customer service expectations set the basis of asset management. Regulatory standards or local political interests typically establish a minimum level. The asset-management system provides a means to protect, maintain, or improve the asset value of a collection system with planned maintenance and repair based on predicted deterioration. In a private or public utility, key information is needed to manage costs through asset-management planning. This information includes conditions and performance of assets; operating costs; financial position including revenues, balance sheet, and cash flow; required and anticipated future levels of service; and methods of measuring and monitoring performance of the system.

The primary goals of wastewater collection systems operation are to prevent public health hazards and to protect the environment by transporting wastewater uninterrupted from its source to the treatment facility. The goal of asset management is to improve and document the decision-making process necessary to protect, maintain, or improve the value of this asset (the collection system) while providing the level of service desired. These issues are described further in each subsequent chapter of this manual. A glossary of terms is included as Table 1.1.

- Chapter 2—System Operations and Maintenance provides an overview of best management practices associated with operating and maintaining a wastewater collection system, including strategies to improve the efficiency and effectiveness of these operations.
- Chapter 3—Information Management describes key information management principles that are critical to wastewater collection system management. It also describes the importance of collecting and using information to make better, more informed decisions to improve system operations.
- Chapter 4—Collection System Assessment and Capital Improvement Planning discusses the process of planning for capital improvements to the system and setting priorities, including activities needed to assess the condition of existing

TABLE 1.1 Glossary of terms.*

Term	Definition
Cleanout	A vertical pipe with a removable cap extending from a sewer service lateral to the surface of the ground. It is used for access to the service lateral for inspection and maintenance.
Collector sewer	Sewers, approximately 200 mm (8 in.) or smaller that provide service to a localized area.
Firm pumping station capacity	Maximum amount of wastewater flow pumped by a pumping station with the largest pump out of service.
Force main	A pressurized line that conveys wastewater from a pumping station.
Gravity or main lines	The largest portion of the wastewater collection system. They use changes in elevation to transport wastewater by gravity between points.
I/I	Inflow and infiltration is the total quantity of water from inflow, infiltration, and rainfall-dependent inflow and infiltration, without distinguishing the source.
Infiltration	The introduction of groundwater to a sanitary sewer system through cracks, pipe joints, manholes, or other system defects.
Inflow	The introduction of extraneous water to a sanitary sewer system by direct or inadvertent connections with stormwater infrastructure, such as gutters and roof drains, uncapped cleanouts, and cross-connections with storm drains. Dry weather and wet-weather surface runoff entering the system through holes in manhole covers or other component will also contribute to inflow.
Interceptor	Large gravity sewer pipes typically having no direct service connections that convey wastewater from trunk sewers to the treatment facility. These were typically constructed at the same time as wastewater treatment plants, thus were designed to intercept wastewater previously flowing directly into receiving waters and convey these flows to the plant.
Level of service	A measure by which utility operators and managers measure the quality and reliability of service or care provided to the customer based on infrastructure design standards, operating procedures, and maintenance frequency.
Lift or pumping station	A mechanical method of conveying wastewater to higher elevations. A lift station is typically distinguished from a pumping station in that the length of the discharge piping from the station is relatively short and intended primarily to lift the wastewater to a higher elevation so it can continue to flow by gravity through sewers. It can be constructed within a reasonable depth from the surface.

* Definitions or terminology used in this manual may differ from those used in local municipal codes or state and federal regulations. Legal definitions of these terms should take precedent when applicable.

Term	Definition
Manhole or junction box	A manhole or junction box provides a connection point for gravity lines, service laterals, or force mains and as an access point for maintenance and repair activities.
Pressure-reducing pumping station	Inline pumping station, typically variable speed and found in a manifold, force-main type of wastewater transmission system designed to reduce pressure upstream of the station.
R value	Used in flow monitoring data analysis or in a hydrologic model to represent the fraction of rainfall in a basin that enters the sewer system as rainfall-derived infiltration and inflow (RDI/I).
RDI/I	Rain-derived (or dependent) infiltration and inflow. It is I/I that occurs as a result of rain events and does not account for groundwater infiltration.
Sanitary sewer overflow	An overflow, spill, or release of wastewater from the wastewater collection and treatment system, including all unpermitted discharges, overflows, and spills. May also refer to releases of wastewater that may not have reached the waters of the United States or, in some interpretations, includes building backups.
Sewer service laterals (or service lines)	A portion of a sanitary sewer conveyance pipe, including that portion in the public right of way, that extends from the collector sewer to the single- or multi-family apartment or other dwelling unit or structure to which wastewater service is or has been provided. May also be referred to as a private lateral if owned by the customer and not by the public utility.
Surcharge condition	Condition that exists when the supply of wastewater is greater than the capacity of the pipes to carry it and the surface of the wastewater in manholes rises to an elevation above the top of the pipe and sewer is under pressure or head, rather than at atmospheric pressure.
Transmission capacity	The capacity of pumping stations and force mains that convey flow to the collection system or treatment plants. This may also refer to the capacity for gravity sewers to convey flow before reaching surcharge conditions.
Trunk sewer	Sewers, approximately 200 mm (8 in.) or larger, that convey wastewater flow from the collector sewers to large pumping stations or the wastewater treatment plant.
Vacuum sewer system	A sewer system that relies on differential pressure between atmospheric pressure and a partial vacuum applied to the piping network to convey wastewater flows. This differential pressure allows a central vacuum station to collect the wastewater of several thousand individual homes, depending on terrain and the local situation. Vacuum sewers take advantage of available natural slope in the terrain and are most economical in flat sandy soils with high groundwater.

assets such as field activities for structural and flow evaluations and computer-based modeling tools for system capacity evaluations.

- Chapter 5—System Design Considerations introduces key design principles and considerations critical to wastewater collection system managers so that life-cycle costs are adequately considered and design decisions consider risk management.

- Chapter 6—Construction Contracting provides the collection system manager with an overview of construction contracting concepts including procurement considerations, legal considerations, and how to manage quality control during construction activities.

- Chapter 7—Public Policy and Community Relations deals with ways to improve communications with collection system customers and other stakeholders with interests in collection system management as well as the development and implementation of public policy and level of service decisions.

- Chapter 8—Budgeting and Financial Planning summarizes key financial considerations for the collection system manager including the basis for generating revenues needed to operate and maintain the system and financial management principles needed to properly account for system depreciation and renewal requirements.

- Chapter 9—Safety, Standard Procedures, Training, and Certifications describes critical processes needed for a safe operating environment for utility employees and customers, and summarizes training and certification programs that will improve understanding and compliance.

- Chapter 10—Emergency Preparedness and Security reviews issues that must be considered to prepare for both manmade and natural emergencies and disasters, including activities that can increase awareness and improve responsiveness of utility personnel during an emergency.

In addition to sewer pipes, pumping stations, and force mains, several other assets must be managed for effective system performance. Examples include human resources that must be managed through proper training, education, and planning and information resources that require data management, quality control, and archiving standards. Other assets include intangibles such as consumer confidence, management integrity, and leadership vision. In many ways, management of these assets is as important as management of physical assets to the long-term viability and success of wastewater collection system programs.

3.2 Capacity, Management, Operations, and Maintenance Program Principles

An excellent example of the industry's increased focus on asset management is the U.S. EPA's development of guidance documents describing the key wastewater collection system programs associated with capacity, management, operations, and maintenance, also known as CMOM programs (U.S. EPA, 2005). The premise of the CMOM program is that when a utility incorporates good business principles into its organization, its wastewater collection system will meet the intended performance objectives resulting in fewer sanitary sewer overflows (SSOs). The CMOM program places the burden of proof on the system owner to demonstrate that SSOs are being prevented to the maximum extent practical by using pipes, pumps, and infrastructure with adequate capacity and by properly managing, operating and maintaining the system,. The CMOM programs promoted by U.S. EPA are consistent with basic asset-management principles as evidenced by the following descriptions of the key CMOM elements:

Capacity program elements promote an understanding of the capacity of system components (gravity sewers, pumping stations, force mains) as designed and under current conditions, tracking how much of that capacity is used, and forecasting when additional capacity will be needed. All of these elements are necessary to support better capital improvements planning. This includes the use of flow monitoring and other evaluation tools to identify where and when capacity is needed to handle current or future increases in wastewater flows and what facilities or system improvements may best meet those needs. In some cases, these analyses may include identifying where infiltration and inflow (I/I) is depleting available system capacity and where the implementation of system rehabilitation to recover that lost capacity may be cost effective.

Management program elements promote efficient and cost-effective decision-making and priority-setting through sound organizational structures, good internal and external education and communications, performance goals/objectives, and proper document control and information management. Good management principles also include having sufficient legal authority, adequate resources, and clear leadership directives to meet program objectives consistently.

Operations programs must promote efficiency and reliability through the development and implementation of standard procedures, training of staff in those procedures, the tracking of system performance through monitoring systems, and the documentation of the location and condition of system components. This includes mapping programs and inspection of new construction and rehabilitation to promote the use of high quality materials and proper installation.

Maintenance programs are the backbone of ensuring consistent system performance through scheduled system cleaning and maintenance activities as well as emergency or reactive maintenance to unanticipated events. This includes the organization and tracking of inventories for materials and equipment.

As mentioned previously, CMOM programs are not a national regulatory requirement at the time of this publication. However, the industry is adopting CMOM principles for several reasons. Many regulatory enforcement actions have required utilities to adopt CMOM programs as terms of compliance with consent decrees or special orders of consent. In addition, several state regulatory agencies with wastewater enforcement authority require CMOM programs through collection system or NPDES permitting requirements. Other communities are adopting CMOM principles because they can improve performance of the wastewater collection system.

This manual can help utilities managers understand and apply asset-management and CMOM principles to improve performance of their systems. With these programs in place, managers can become a leader in better management of infrastructure assets to improve the quality of life and environmental sustainability of communities.

4.0 REFERENCES

Albee, S. (2005) Finding a Pathway for Sustainable Water and Wastewater Services. *Underground Infrastructure Management*, November/December 2005, Benjamin Media, Inc.: Peninsula, Ohio.

Federal Highway Administration; American Association of State Highway and Transportation Officials (1996) *Asset Management: Advancing the State of the Art Into the 21st Century Through Public-Private Dialogue;* Federal Highway Administration: Washington, D.C.

National Association of Clean Water Agencies; Association of Metropolitan Water Agencies; Water Environment Federation (2007) *Implementing Asset*

Management: A Practical Guide; National Association of Clean Water Agencies: Washington, D.C.

U.S. Environmental Protection Agency Office of Enforcement and Compliance Assurance (2005) *Guide for Evaluating Capacity, Management, Operations, and Maintenance (CMOM) Programs at Sanitary Sewer Collection Systems;* U.S. Environmental Protection Agency: Washington, D.C.

5.0 SUGGESTED READINGS

U.S. Environmental Protection Agency Office of Water (2004) *Report to Congress: Impacts and Controls on CSOs and SSOs;* EPA-833-R-04-001; U.S. Environmental Protection Agency: Washington, D.C.

Water Environment Federation; American Society of Civil Engineers (2008) *Existing Sewer Evaluation and Rehabilitation,* 3rd ed; Manual of Practice No. FD-6; ASCE Manuals and Reports on Engineering Practice No. 62; McGraw-Hill: New York.

Chapter 2

System Operations and Maintenance

1.0	INTRODUCTION	14	4.5	System Problems and Overflow Response	31
2.0	PROGRAM OBJECTIVES	14	5.0	PUMPING STATION OPERATIONS AND MAINTENANCE PROGRAMS	34
3.0	ESTABLISHING OR EVALUATING AN OPERATIONS AND MAINTENANCE PROGRAM	15			
	3.1 System Asset Characteristics	15		5.1 Increasing Importance of Pumping Station Maintenance	36
	3.2 Performance History	16		5.2 Types of Pumping Station Maintenance	37
	3.3 Agency Resources	16			
	3.4 Customer Expectations/ Complaints	17	6.0	PUMPING STATION OPERATIONS AND MAINTENANCE STAFFING	40
	3.5 Community Interests	18			
	3.6 Regulatory Resources	19			
4.0	PROGRAM COMPONENTS	19		6.1 Pumping Station Operations Staffing	42
	4.1 Maintenance	23		6.2 Pumping Station Maintenance Staffing	43
	4.2 Mapping Use	27			
	4.3 Directed Maintenance Programs	27	7.0	OPERATIONS AND MAINTENANCE FACILITIES AND EQUIPMENT	44
	4.4 Application of Technology to Support Operations and Maintenance	29	8.0	REFERENCES	46

1.0 INTRODUCTION

Managing an efficient collection system operations and maintenance (O & M) program requires reliable resources, dynamic planning, adequate budgets, and frequent performance reviews. For success, managers also need to consider stakeholder interests, system composition and complexity, system performance history, system maintenance protocols, and regulatory requirements. This chapter reviews various considerations in establishing a new program or evaluating an existing one.

2.0 PROGRAM OBJECTIVES

When establishing or reevaluating any O & M program it is important to have a clearly defined objectives or purpose. The main purpose of a collection system is to provide a means to collect and transport wastewater (sanitary and storm) away from its source while protecting public health and safety and the environment. This objective is accomplished by either a dedicated storm sewer with a separate dedicated sanitary sewer or with a combined sewer that conveys both stormwater and sanitary wastewater flows within the same system. The system operator is responsible for where, when, and how this is accomplished.

Stakeholders—which include rate payers, regulatory agencies, other utility and service agencies, the media, and employees—will have varying views, opinions, and interests in a collection system program. These considerations can result in a more manageable and successful program through building trust and credibility, defining responsibilities, and enabling participation. It's also important to communicate an understandable, consistent message to stakeholders.

Managers need to take into consideration several factors when reviewing or establishing new processes in collection system O & M program:

- Regulatory requirements;
- System operation and performance goals based on regulatory and/or agency requirements;
- How rates are used to accomplish requirements and goals;
- If and how program affects and is influenced by other public or private entity programs;
- Responsibility of the sewer agency to provide service to its customers;
- Responsibility of the sewer customer for receiving service from the sewer agency;

- How a customer has access to sewer service through permitting and related fees;
- Customer communications; and
- Responsibility to protect the environment in all activities.

3.0 ESTABLISHING OR EVALUATING AN OPERATIONS AND MAINTENANCE PROGRAM

Although there are some basic regulatory guidelines for system performance, establishing, organizing, and reevaluating a program may save the operator time and money. Data should be collected for several parameters that affect O & M: system asset characteristics, asset maintenance and operational history, asset manufacturers' recommendations, utility resources, customer expectations/complaints, community interests, long-term planning, and regulatory specifics. Each system may have unique parameters to consider. A practical approach that helps a system meet regulatory and customer expectation is recommended.

Following are examples of how data collection and data use affects an O & M program.

3.1 System Asset Characteristics

System asset characteristics include physical attributes such as size, age, material type, sewer type, location, depth, access, slope, flows, and content.

For example,

- Trunkline A:
    - A 450-mm (18-in.) diameter gravity sewer main,
    - 10 years old,
    - Constructed of polyvinyl chloride pipe,
    - Sanitary sewer,
    - Inside a right-of-way,
    - With better than minimum slope,
    - Few physical connections, and
    - Quarter to half-pipe diurnals.
- Trunkline B:
    - Approximately 450-mm (18-in.) diameter gravity sewer,
    - 70 years old,

- Constructed of concrete,
- Combined sewer,
- In easement along creek bank,
- Minimum or below-minimum slope,
- Diurnals ranging from zero flow to half-pipe, and
- Multiple commercial connections including restaurants (grease).

In this example, an operator may consider inspecting and cleaning trunkline B on a more frequent return cycle than trunkline A. Primary reasons for this approach would be type of material; joint condition; limited access during wet weather; typical low-flow conditions and slope, which creates the potential for debris collection; and grease buildup because of the restaurants (see the discussion on fats, roots, oils, and grease programs later in this chapter). Over time, field maintenance crew experience, observation of sewer system use, maintenance and corrective action records, or increases in flow may result in a different return cycle than originally determined. Further discussion of information management and using data to support better O & M decisions is provided in Chapter 3 of this manual.

3.2 Performance History

Depending on the nature and cause of a collection system problem, an operator may determine that a location requires a more frequent return cycle for evaluation. Common causes of problems include grease, roots, debris, pipe failures, and overflows. Defects that cause mainline failure are best remedied through permanent correction, which might not always be immediately practical or possible. An example would be a sewer main through a central business district that requires heavy, semi-annual cleaning. This sewer main may be identified as not requiring capital outlay for repair or replacement because of its location. It may remain on a cleaning and inspection cycle until conditions change, although this approach may not be standard for the rest of the system. Operator and staff will be able to make more informed decisions as they become familiar with the performance of the collection system through data and performance measure reviews.

3.3 Agency Resources

Depending on system size and budget, operators have many considerations and options concerning resources to accomplish program goals, including staffing and equipment needs.

Program objectives need to be weighed against staffing ability. Some references may recommend staffing levels based on system size, but technology use may affect those numbers. Determining an initial return cycle of the system for preventive maintenance and the ability to respond to emergencies can be based on collected data to determine if staffing levels are adequate. Although attrition is reducing the number of knowledgeable staff, contract services have become more readily available. Periodic review of contract service availability and cost can be good information for the operator to compile and to compare to in-house staffing costs.

Program objectives also need to be weighed against equipment availability and the return frequency of pipeline and facility review or cleaning. Costs per unit of work on a regional or units-per-man-hour basis for cleaning and televising are available through industry organizations or third-party vendors and contractors.

3.4 Customer Expectations/Complaints

Public awareness is becoming more of a consideration as it relates to program objectives and deliverables because of the immediate availability of information. This means that the operator must use judgment on how information is communicated to the public to ensure that the definition and execution of the program protects the agency from potential liability.

An example of operator judgment would be noting and reporting a sewer manhole overflow in a remote portion of the collection system. If the overflow extends outside of the easement or affects a receiving stream, then public notice would be appropriate. If it is contained in the immediate area, does not affect open land or receiving streams, and does not interrupt service to customers, then notification to the regulatory agency and corrective action may be enough. Information for events such as these could be posted on Web sites or on required regulatory reports with details shared upon request of customers, rather than unnecessarily alarming the public.

As indicated in other areas of this document, stakeholders can include system users, land developers, regulatory agents, and concerned citizens. Their perspective and interests can affect an operator's program. Basic information of interest to most customer types includes capacity of the collection system, appropriate use of the system, and the customer's responsibility in helping to maintain it.

As development occurs, an operator must make decisions about whether an existing system can accept the flow from a new connection or if an extension of the system is possible. The answers to these issues depend on the capacity of the system and availability of financing for improvements needed to serve development. New service agreements are recommended to be executed in writing and made publicly available to eliminate the potential for misunderstandings. Further discussion of capacity assurance planning is provided in Chapter 4 of this manual; system financing options are discussed in Chapter 8.

It is recommended that there be stated conditions concerning use of the collection system. This is typically accomplished through a sewer use ordinance adopted by the local governing body. These documents typically correlate with other local code involving permitting, construction, inspection, and wastewater acceptance requirements. Specific conditions can include industrial or environmental compliance concerning pretreatment, which ultimately will require periodic testing and enforcement for noncompliance. Conditions also may include methods of connection and allowable discharges. For instance, roof drains and sump pumps would not be acceptable in a separated sewer system but may be in a combined sewer system. Also yard waste, dirt, rocks, construction waste, rags, personal hygiene products, and household grease are examples of items that should not be disposed of in a sewer collection system because of risk of line stoppages and overflows. Information such as this should be communicated to proactively educate the customer.

Just as the system operator has responsibility for maintaining and operating a collection system, so too should the customer have responsibility in its use and maintenance. In addition to safe disposal, the customer also should be responsible for the condition of their service line, eliminating the potential for root intrusion and capacity-depleting groundwater infiltration. The system operator should make this clear in local code, upon connection for service to the system, and through various agency communications. Sometimes, financial hardship may affect a customer's ability to make required corrections to their service line. The utility will need to determine how to address this and whether system expenditures on private portions of the collection system will yield a return worth pursuing. Many locations have determined that addressing private laterals and eliminating unapproved connections (roof drains, sump pumps, etc.) has contributed to removal of extraneous flow and its related non-revenue effect to the system. Some systems have established loan or grant programs for qualifying customers.

3.5 Community Interests

The development of any area is important to economic growth and quality of life. Operators must coordinate with development interests in use of existing or construction of new infrastructure. Receiving sewers, pumping stations, and treatment facilities must be considered when enabling new connections. Type of use and discharge are other factors to consider. A master sewer plan based on existing and potential land use is a good tool to facilitate discussion about development and the interests of the community. Typical master plans project future conditions for 25 years or more and should be reviewed and updated periodically. Master planning and capital improvement planning is discussed in Chapter 4 of this manual.

3.6 Regulatory Resources

Interpretation of regulations by local agencies is critical to funding, staffing, equipment, and overall program objectives. Regulatory agencies can serve as a great resource for information and as an advocate with other stakeholders. As with any relationship, trust is earned through communication and demonstration of responsibility and accountability. A collection system operator may achieve this trust through timely, credible, and dependable reporting. Operators can further develop relationships by working on industry activities that enable cross-agency interaction and through good system management to ensure few incidences or failures occur.

4.0 PROGRAM COMPONENTS

Stakeholders may find useful several components of the system used by operators. These items should be designed accordingly. Sensitive or secure information should also be taken into consideration as it relates to privacy, terrorism, and other misuse. Program components that stakeholders may find useful are outlined below.

Service maps may include spatially related service area as constructed, topographical information, street addresses, and customer information. This level of detail assists in planning, service availability requests, and in field responses to complaints.

Organizational charts of utility personnel by function define who is responsible for what activity and how to contact them.

Asset data may include size, material type, age, and depth and may be contained in a database or mapping system. Typical database formats include a computerized maintenance and management system (CMMS) for asset characteristics and history or a geographical information system (GIS) for spatial relationship representation. Information may be used in planning, system analysis, budgeting, service availability reviews, and field responses to complaints.

Several types of data might be collected for a sewer pipe:

- Size (diameter);
- Length (miles, footage, and laying length);
- Type (gravity, force main, collector, interceptor, lateral, trunk, siphon, or pipe bridge);
- Material (vitrified clay pipe, concrete, polyvinyl chlorine, cast iron pipe, ductile iron pipe, or high density polyethylene);
- Age (install date);
- Depth (cover for excavation and overburden evaluation);
- Location (right-of-way, easement, or private property);
- Service connections (tee stations, depth, length, material size); and
- Historic condition—logs, televising, maintenance and repair records (work orders).

Information that might be collected for manholes (manways) include:

- Type (concentric, eccentric, chimney, tee, catch basin, regulator, vault);
- Location (right-of-way, easement, or private property);
- Size and depth;
- Casting type and diameter;
- Casting and invert elevations; and
- Historic condition.

Various pieces of data should be collected for pumping stations (for more guidelines, refer to the Pumping Station Operations and Maintenance section):

- Location (street address or other geo-coding for mapping);
- Type (centrifugal, ejector, grinder, submersible);
- Schematics (piping, electrical), which assists in understanding typical operation and in modifications during emergency or maintenance activities;
- Controls (what controls what);
- Equipment (install and service dates, maintenance records, warranties);
- Backup power (remote or onsite);

- Valves (isolation, purge, and air relief);
- Force main (size, material, length, terminus);
- Metering (type, location, reading units); and
- Historic condition.

Standard operating procedures (standardized format throughout agency) can be used as a training tool and reference by employees and other stakeholders. Work-process flowcharts (standardized format throughout agency) also can serve as a training tool and reference. Several other plans are valuable:

- Operations and maintenance manuals—assists as a training and reference tool by employees and other stakeholders;
- Master plans—multiyear outline of system expansion or modification that meets specific operational projections;
- Policies—application guidelines of regulation or operational practices); assists as a educational tool for customers and as a training and reference tool for employees;

Special programs—regulatory or capital-driven projects—with specific regulatory or operational objectives should also be documents. Examples of special programs may include:

- Capacity, management, operation, maintenance (CMOM)—a U.S. Environmental Protection Agency (U.S. EPA) program that establishes the basis for separated sewer collection systems for better operation using measures and goals as they relate to total system performance and improvement.
- Long term control plan (LTCP)—U.S. EPA program that establishes requirement for communities with combined sewer systems; shows progress toward standards consistent with the Clean Water Act by demonstrating adherence to water quality and improvements as required and affordable.
- Nine minimum controls—U.S. EPA program that establishes the basis for the control and management of a combined sewer system. Part of the LTCP.
- Rehabilitation—various operational and capital activities that reestablish system integrity, performance and reliability.
- Expansion—various capital activities that result in an expansion of service area. Typically part of a master plan or 201 facilities plan.

The customer information system (CIS) is composed of a database for connections, metering, billing, and contracts. Information should also be gathered on employees, performance evaluations, and planning as outlined below.

- Full-time equivalent employees:
  - Number—how many directly assigned to collection system O & M;
  - Function—for example, vacuum, jet (flushing), combination (jet and vacuum), trunkline, manhole, maintenance, or repair;
  - Location—assigned portion or zone of system;
  - Shift (day and time);
  - Certifications and/or license designations;
  - Hire date and/or experience level; and
  - Special skills.
- Part-time employees:
  - Number—how many directly assigned to collection system O & M;
  - Function—for example, vacuum, jet, combination, trunkline, manhole, maintenance, or repair;
  - Location—assigned portion or zone of infrastructure;
  - Shift—day and time;
  - Certifications and/or license designations; and
  - Hire date and/or experience level.
- Emergency response:
  - Current on-call listing of in-house resources—should be updated every six months and based on guidelines for reporting, substitution, compensation, and disciplinary action for nonresponse; and
  - Contract services—current contact information.

Equipment itemizations and categorizations contained in a CMMS can be a valuable reference tool to help track availability, schedule maintenance and repairs, and prepare capital budgets. The CMMS can also record installation, warranty, and maintenance history. Units may include jets, vacuums, combination, closed-circuit television (CCTV), jet nozzles, saws, rodding machines, bucket machines, boom truck, and safety equipment. Safety equipment should include Occupational Safety and Health Administration required personal protective equipment, gas monitors, ladders, picks, and shovels. Below are details that should be kept for each unit:

- Number of units,
- Uses of units,
- Training types and schedules,
- List of in-house, certified equipment operators,
- Emergency contracts with contact information, and
- Inventory or vendor access to spare parts.

4.1 Maintenance

As stated previously, the system operator must balance an efficient and effective maintenance program for the collection system with safety, regulatory compliance, budgetary limitations, customer expectation, and resources. This is a dynamic process as the system ages, as regulations change, and as the system is expanded.

In the past, many systems and components were run to failure and simply repaired or replaced as quickly as possible. More recently, redundant or dual systems and routine maintenance have been incorporated where required or practical. Because continual service is a goal of collection system operations, maintenance is geared to prevention and prediction. As recommended above, detailed records of materials, warranties, installation dates, and historical O & M records help to establish effective maintenance schedules. This more-calculated approach ultimately will save time, money, and service interruptions.

Several activities, equipment types, and processes must be considered in planning for system maintenance. These activities include:

- Bypass pumping—approximately 75 to 100 mm (3 or 4 in.) submersible type with reliable piping;
- Emergency power— portable generators to operate pumps and for emergency task lighting;
- Hauling—tankers to transport collected flow;
- Disinfection— sodium hypochlorite to treat spills or overflows;
- Construction of repairs— construction equipment and materials (backhoe, tandems, dump trucks, trench boxes, pipe, fittings, and gravel backfill; and
- Bypass discharge—to route bypass pumping downstream or to another sub-basin point.

The type of communication depends on whether it the event is scheduled or an emergency, such as an overflow because of system failure. A notice to both the appropriate regulatory agency and affected customers may be necessary.

For scheduled activities, depending on location and effect on customer, notice may be helpful advising of various specifics:

- Activity—repair and/or heavy cleaning (removal of debris);
- Timeframe—start time and estimated completion time;
- Service interruption—if system use is affected;
- Agency contact information for any questions or followup;
- Method of communication—depending on location and customer effect, notice may include one or more of the following:
  - Posted signs advising that work is being performed and to take caution if in or around roadways.
  - Door hanger.
  - Out-bound call.
  - Post-event notices—depending on cause of the repair or maintenance activity, this notice may help to educate customers of the hazards of roots, extraneous flow, grease, and other materials and to reduce the potential for future events.

Emergency activity may require use any of the previously mentioned communication tools and any or all of the following:

- Notice of overflow—document to communicate to regulatory agencies about the event, its environmental effects, and corrective action taken and
- Site signage—depending on severity and location, onsite signage to alert the public to avoid the area.

There are three types of maintenance: corrective, preventative, and predictive. Each has a set of suggested activities and considerations as outlined below.

Corrective maintenance either occurs during an emergency or is scheduled, resulting in repair or replacement of an obsolete or defective component. Depending on component type, temporary measures may be required to enable service and minimize or eliminate environmental effects. When responding to an emergency or planning scheduled corrective maintenance, the supervisor should consider what supporting activities may be required to make the repair and still meet system objectives.

Preventive maintenance is performed on a predetermined schedule on selected infrastructure. This type of maintenance is performed to prevent failure, reach, or prolong useful life and to ensure that operating components meet system goals. As detailed previously, good records and system performance knowledge assist in establishing efficient, effective preventive maintenance and can help involve maintenance staff in determination and scheduling of work.

Preventive maintenance can address both general collection system function and recurring problems where rehabilitation or reconfiguration is not immediately feasible. An example of this latter group is cleaning grease-prone areas. Preventive maintenance can include CCTV for visual condition assessment; jet/vacuuming (light cleaning); rodding/bucketing (heavy cleaning); and trunkline walking. Although condition-assessment activities such as CCTV are often conducted by O & M staff, the information collected is useful for many other aspects of system planning and asset management as further described in Chapters 3 and 4.

All these activities can be performed during dry and wet weather; although there are considerations for each. Activities for dry weather include

- Closed-circuit television—to detect pipe condition, service connection locations or conditions, and manhole condition assessment.
- Routine cleaning—chronic grease, root, or debris settlement (dips in line) creating stoppages, backups, and overflows.
- Trunkline walking (surface inspection)—to check manhole and casting integrity and condition; any visible surface area depressions over sewer main (indicating potential failure); service line cleanout caps; flow characteristics through manhole and the presence of any collected solids; and comparison of field observed connections to CIS. Pole-mounted cameras may also be used to photograph manhole condition and features and to determine pipeline characteristics. Some pole-mounted cameras use zoom lenses to allow inspections up to 12 m (40 ft) upstream and downstream of the manhole as long as there are no obstructions in the sewer line.
- Smoke testing—to discover inflow and infiltration points including pipe-joint leaks, storm-sewer cross connections, missing service line cleanout caps, connected roof drains, sump pumps, or other illegal connections.

Activities for wet weather include:

- Closed-circuit television—to detect inflow and infiltration points during or immediately following rain events (may require line plugging to control upstream flow for optimum viewing).

- Trunkline walking—direct visual inspection to determine sewer segments of wet weather affect sources through observed and/or measured flowrates for further investigation (CCTV or smoke testing).

Predictive maintenance requires using knowledge of the system, including past performance; types of connections; physical characteristics of the sewers (size, grade, material, and age); hydrologic conditions (recent rainfall and groundwater elevations); and common sense. The wastewater collection system operator can predict and schedule preventive maintenance in practical ways. By using this approach and tracking specific system measures, the operator will be better able to represent and defend their programs.

Specific system measures can include the number of overflows per unit of distance. This measure is considered representative of maintenance and system operation success (the lower the number the more successful the program; should trend downward over time).

Measures also can include the causes of overflow such as rain, grease, roots, debris, collapsed pipe, or vandalism. Again, there should be a trend downward as maintenance programs are defined and executed.

Rainfall-induced overflows may be tracked by number of events, duration, and volume; and by storm event, time of year, and groundwater table. Data used to define rehabilitation needs on a sub-basin or system-wide approach. There are many variables to this issue and many resources available from U.S. EPA and the Water Environment Federation, including *Existing Sewer Evaluation and Rehabilitation* (WEF and ASCE, 2008).

Grease-induced overflows may be tracked by number of events and location. There should be a downward trend in the number of events as system maintenance, customer education, and grease management programs (interceptor inspections and enforcement) increase.

There are two primary ways to measure overflow:

- Controlled method—volumes measured through a direct conduit (pipe, structure) by weir or flow monitor; and
- Estimated method—volumes based on observed start time and end times incorporating surface area, depth of overflow, and rate of discharge based on height of discharge from manhole or orifice compared to drawings or pictures with predetermined discharge rates.

Depending on the pervious/impervious nature of the discharge area, the latter method is challenging when considering ground absorption and irregular patterns of overflow. It can, however, serve as a reasonable approach for reporting. Containment and retrieval of wastewater during spills can also assist in estimating overflow volumes.

4.2 Mapping Use

Good system mapping is critical for an effective O & M program and to provide a geographic basis for implementing an information management system as described in Chapter 3. Mapping enables spatial representation of infrastructure for planning, field locating, and maintaining infrastructure, which means that accuracy is important.

Maps can serve as a tool to visually demonstrate completed work and for determining future work as an agency works through their system. This is also important when using a one-call center for infrastructure field locating at the work site. This visual and spatially related method is better than lists or tables. Mapping work activities also is helpful in demonstrating the magnitude and effect of these activities to stakeholders (regulatory, development, ratepayers) when representing production goals, reducing liability, budgeting, and regulatory reporting.

Mapping format should be carefully considered. Although hardcopy maps work, electronic systems are recommended because of ease of editing (extension or field correction of misrepresented mapping data); sharing (providing maps to other agencies or entities to overlay with proposed designs); and presenting special data overlays (demonstrating locations with documented defects for rehabilitation or repair projects). There are a variety of mapping programs available.

Using a map to represent corrective, preventive, and predictive maintenance goals and accomplishments is a tool the system operator can use to educate all, including management, employees, regulatory agencies, and customers.

4.3 Directed Maintenance Programs

Besides standard maintenance and repair processes, the system operator may incorporate specific programs to overall system management. The programs are summarized in this section.

A fats, roots, oils, and grease program aims to manage situations that can result in line stoppage and potential overflows. Such a program may include many specific tasks such as a grease-interceptor inspection. This would require documented, periodic inspections of commercial service lines requiring grease interceptor installation and maintenance. Targeted collection system monitoring would include CCTV of existing, chronic or potential system buildup locations (commercial and residential) for subsequent corrective/preventive cleaning.

Regardless of system components, education is vital to success. This might consist of letters of notice or advisement of responsible system use in regard to discharge of fats, oils, and grease and the control of roots through private service lines. The ability to capture still photographs of service-line conditions demonstrating roots or grease buildup is a helpful tool for informing customers and enforcing compliance.

Enforcement helps to establish new or to revise existing local code to enhance the collection system's ability to enforce non-compliance with service connectivity requirements. Codes needs to be shared with customers through direct mail, Web site posting, or bill inserts. Enforcement action options include fines, fees, agency cost-recovery, and disconnection of potable water.

Depending on severity, source, and effect, the wastewater collection system operator may need to mechanically or chemically control roots. The operator also may need to repair, replace, or rehabilitate the mainline sewer. Less severe locations need to be placed on a periodic inspection and cleaning schedule and monitored for more complete correction. There are a variety of options to control roots and the operator needs to know availability, cost, and success rate of each to determine which to use in their system. Mechanical rodding machines or special hydraulic cutting tools or flushing machines, which use intermediary or full-pipe-size tools, are the most common means of real-time root control. In some areas of the system, long-term root control may be accomplished through use of chemical applications. These may be high-traffic areas that endanger personnel on the street, easement areas that are difficult to access, or areas where roots prove are difficult to remove with conventional maintenance equipment. Situational determinations of root severity, root source, growth rate, and location of root intruded sewer segments in relation to the water table need to be considered in any root-control determinations.

A rehabilitation program focuses on correcting system defects that allow unmanageable levels of inflow and/or infiltration or that constitute an unacceptable structural or maintenance problem. General structural integrity and flow characteristics are also considerations. Like other programs, rehabilitation approaches and methods depend on physical considerations such as:

- Severity of pipeline condition (hairline cracks, missing pipe, joint leaks, structural integrity, etc.);
- Location of pipe (easement or right-of-way, central business district or residential);
- Resultant surface restoration effect; and
- Pipeline location in relation to watershed (upper reaches, mid or lower areas).

As mentioned previously, mapping defects is an essential tool for understanding the extent of system problems and determining the best method of correction.

The U.S. EPA Web site outlines programs to standardize certain aspects of system O & M using an organized and systematic approach, including CMOM, LTCP, and nine minimum controls.

The CMOM program's objective is to provide a guide and a checklist for collection system operators to evaluate O & M processes in coordination with U.S. EPA and state regulatory agencies.

The LTCP program's objective is to assist the system operator in development of a plan to address combined sewer overflows in relation to the U.S. EPA's combined sewer overflow (CSO) control policy.

The nine minimum controls program's objective is to guide the collection system operator in reducing CSOs and their effects on receiving water quality.

4.4 Application of Technology to Support Operations and Maintenance

As with many industries, collection system O & M has been affected by an increasing level of administrative duties. These duties include documentation and tracking of system performance and status relative to condition, measure flows, system problem response time, regulatory notification and reporting, customer notification and reporting, environmental monitoring, and mandated program metrics. As these duties have been defined and have evolved and become standardized, technology has emerged that can provide assistance. One important consideration in the use of any technology is its compatibility and capacity to interface with other organizational technologies. This section reviews examples of current technologies and their general application and use in collection system O & M.

Geographic information system technology is an electronic spatial representation of a collection system in relation to other subterranean and topographical features. There are many software products available for this purpose depending on specific needs of the system. Use of this technology enables field response to problems, allows for real-time updating of field-discovered discrepancies in configuration, provides specific program applications such as pipe defect mapping for rehabilitation, and supports long-range planning and development through projections and scenarios.

Computerized maintenance management systems can be used to track asset characteristics and maintenance and performance history. It also can be used to schedule preventive maintenance and as tool for condition assessment (for example, rain-induced sewer stoppages for the last five years) when interfaced with a mapping tool (GIS). Use of this type of system involves extraction of collected data reported on CMMS work orders from field activities by asset or activity type. Completed field activity is documented and reported back into the database for archival and retrieval purposes.

A CIS is a database that can be used as a customer account program for billing and collections. It can also be used with CMMS and GIS for service-connection reconciliation and enforcement by identifying responsible party of record on the connection account.

Flow monitoring technology can be used to measure flow characteristics over time and under various conditions (rain, water table, pipe condition, and pre- and post-rehabilitation). Depending on system size and affordability, it may be beneficial to have permanently installed sites supplemented by seasonal temporary sites. There are also many methods of securing flow readings as basic as manual measurement through graduated rods inserted in flow, utilization of weirs, and more sophisticated electronics that can measure depth, velocity, and pressures.

Through these measurements, an operator can collect information ranging from a single, instantaneous reading through continuous readings at incremental timeframes (seconds). Information gathered through flow monitoring can help focus investigative measures (CCTV, trunkline walking, smoke testing, etc.). It can also be used for capacity management and allocation and calibration of system hydraulic models as described in Chapter 4.

Closed-circuit television enables the ability to "see" pipeline condition and performance. Use of this technology depends on system size, pipe diameters, system age, and wet-weather performance. As with any of the technologies, the system operator must determine the cost effectiveness of developing an in-house ability versus getting third-party support.

Liability is another consideration in using technology. Having a history of pipe condition, system complaints, and corrective action on record may assist in determining necessary corrective action. It may also protect the agency from potential personal and property damages incurred as a result of system blockage or failure and overflow.

There are a variety of CCTV systems with many features for operation and data collection. A variety of camera types are available that are suitable for a range of system pipe diameters and data needs. It is also important to consider needs for interface or integration with other software. Industry recommended standards for pipeline defect coding used in condition assessment must also be taken into account.

Mobile dispatch technology assists in the assignment of both routine and emergency work orders. Scheduled or routine work may also be preplanned and ready for execution by a specific crew at a specific time. An automated dispatch system can improve response time to system emergencies, reduce, or eliminate hardcopy, work-order handling, and enable a more fluid and equitable distribution of work. Systems of this type also may enable electronic entry of field work performed and automatically update agency CMMS. Having this remote capability also eliminates delayed data entry and potential mistakes through a second party. Another feature may include an automated vehicle locator, which assists in dispatching the nearest resource to a system emergency.

Hydraulic modeling technology enables the simulation and prediction of certain system performance outcomes under various operating scenarios. Hydraulic models are typically built from as-constructed plans and calibrated through system flow monitoring. Depending on the type of model used, a system operator may use this tool to:

- Focus investigatory efforts (preventive maintenance) as a result of projected problems;
- Plan for upgrade because of capacity limitations;
- Develop corrective actions (replacement or rehabilitation);
- Develop long range plans for expansion; and
- Allocate capacity for land development.

Several other technologies can assist in the O & M of a collection system:

- Remote sampling for waste content or tracing;
- Surface mounted cameras for remote observance;
- Rain gauges to report and record wet weather activities; and
- Various sensors detecting manhole surcharge activities.

4.5 System Problems and Overflow Response

As with any dynamic system, a collection system will experience problems and failures. As previously stated, an objective of any O & M program is to reduce or eliminate this potential. Any approach should include steps to contain overflow, control the effects of the defect, and then correct the problem. Some sites may not require any containment or control but will simply require correction. It is also important that the utility have equipment that can access the entire system, including infrastructure in easements and roadways. Following are examples of system problems and handling considerations.

Stoppage and overflows problems can range from a partial to complete stoppage and may result in either a surcharge of sewer in the collection system or an overflow. Unless direct observation or an alarm at a pumping facilities or flow monitoring location alerts an operator, customer complaints of slow service, a sewer backup, or an overflow typically lead to discovery of a surcharge or overflow. This includes sanitary waste solids around service line cleanout assemblies, manhole structures, or outfalls. The response to an identified sanitary sewer overflow should be outlined in a clearly documented sewer overflow response plan. The content of such a plan is dependent on system resources and permit and regulatory requirements. Typically, response to a system stoppage involves several steps as outlined below.

The first step is to respond to a reported problem. The agency first respondent may be an investigator, sewer cleaning crew, or CCTV crew. The respondent reviews agency system performance condition through visual inspection of flow characteristics to determine if the problem is with the public main or on the private service line. Depending on respondent type, tools used for preliminary inspection may include:

- System map;
- Manhole pull, pick, or hook;
- Hand-held lamp;
- Jet, vacuum, combination cleaning truck or rodding machine; and
- Closed-circuit television truck.

Depending on the wastewater treatment plant's approach to complaints, CCTV equipment can be used to review complaints to determine if the problem is on the public or private portion of system. If it's on the public portion, CCTV may help determine corrective action (cleaning or repair) or can be used for follow-up to any corrective action.

The next step is to determine and execute corrective action. If the public sewer main is demonstrating surcharge or overflow conditions, then corrective action needs to be taken. This is typically accomplished through one of more of the activities:

- Jetting—hydraulic cleaning;
- Vacuuming—extraction of material;

- Mechanical cleaning rods—physical force to dislodge heavy debris or wedged material;
- Buckets—to remove heavy debris that cannot be jet or vacuum removed;
- Closed-circuit television—to determine pipe condition or potential source of blockage material; and
- Repair equipment—excavation equipment and necessary materials.

It is recommended that any material causing the blockage be removed at a downstream manhole if possible to:

- Eliminate the potential for blockage downstream.
- Enable system flow characteristic assessment because of blockage content (grease, rocks, gravel, construction material, etc.).
- Evaluate potential for liability in the event of personal or property damage loss (source of material from restaurants, industry, development, etc.).

Once corrective action has been taken, the next step is to contain the overflow. The agency respondent may need to contain and disinfect the overflowed wastewater depending on the amount, its location, and the accessibility and topographical condition of the site. The operator also will need to consider any environmental effects that may have occurred. Containment and disinfection can involve pumping, booms or sandbags, temporary storage, and mobile tankage. If containment is necessary, it is recommended that it be performed concurrent with efforts to eliminate the cause of the blockage or overflow.

Next, the spill area must be isolated. Depending on location and potential for public contact, various means can be used to reduce or eliminate the risk. Tools can include barricades, fencing, tape, utility staff, and police.

Once the spill is isolated, notification of the event to the appropriate stakeholders needs to occur. Depending on the severity of overflow, effect on the environment, and proximity to populated areas or receiving streams, the extent and method of communication may vary as previously described. The event also needs to be documented properly as described earlier. Documentation of corrective action is a matter of record for reference in O & M consideration and liability considerations.

The next step is to create a corrective action plan. System problems and response and corrective action are useful information in determining future actions that may include customer notification of use of the collection system, agency emergency preparedness, resource evaluation, and liability considerations.

Finally, any potential odor problems need to be addressed. Potential causes of odors and corrective actions are summarized below.

- Line stoppage—clean and televise for correction.
- Line break—televise and repair.
- System hydraulics and/or hydrogen sulfide—poor design including slopes that do not enable flowrates of between 0.6 to 1.5 m/s (2 to 5 ft/sec); elevation differentials in manholes that cause splashing and the release of gases; terminus of force mains that do not consider hydraulic transitions to gravity flow; and long detention times in wet-wells and/or force mains.
- Nonexistent or poorly managed pretreatment programs for industrial connections—a program and legislative ordinances for enforcement including monitoring and penalties for noncompliance should be established.
- Vent pipes— check location in relation to inhabited or densely populated areas. Relocate or eliminate if possible or consider air filters or deodorant blocks.
- Air-relief valves—check function and condition of valves for proper O & M. Make any necessary repairs.
- Missing manhole covers—locate and repair or replace as needed.
- Open grating (in combined sewer systems)—clean and review long-term options if necessary, including a routine cleaning schedule, air filters, deodorant blocks, and elimination or relocation of grating.
- Defective internal plumbing—educate customers on maintenance of p-traps, wax rings, vent assemblies, floor drains, sump pumps, and cleanout caps. Smoke test if allowable or practical with notice to affected customers.

5.0 PUMPING STATION OPERATIONS AND MAINTENANCE PROGRAMS

Wastewater pumping stations are used when the topography of the service area does not allow wastewater to be conveyed by gravity or where there are other advantages to conveying wastewater via a pressurized system, such as pipeline routing considerations, limiting the depth and width of pipeline construction, or cost. Pumping stations are sometimes referred to as lift stations, typically when force mains are short and the primary purpose of the pumping station is to lift wastewater to a higher elevation so that it can be conveyed by gravity without pipes getting too deep. Pumping stations may also be used in systems with manifolded force mains where significant portions of the pipeline conveyance system operate under pressure to transfer wastewater to the treatment plant.

In addition to representing a significant initial capital investment, pumping stations are more expensive to operate and maintain than gravity sewer systems. Although many of the issues covered in this chapter apply to pumping stations, adequate resources and maintenance programs for both the piping and pumping facility must be maintained to protect the total capital investment. Special skills are necessary for O & M of pumping station equipment. Design, construction, and O & M of pumping stations have changed significantly as a result of new regulatory requirements, environmental and liability issues, increasing size, and safety requirements.

Pumping station failure may result in damage to equipment or the environment or endanger public health. The goal of pumping station O & M is increased reliability and reduced risk of failure. An effective O & M program is a key element in achieving this goal. Proper design, construction, and operator training are also important.

The management of collection systems and pumping stations has become more critical as technology has changed. This section, rather than dealing with specific maintenance tasks for individual pieces of equipment, provides guidelines for a comprehensive pumping station O & M program.

Variations in equipment type, configuration, and geographical factors determine pumping station design and O & M requirements for different systems. Size also plays a role because large station failures tend to have more catastrophic results. Smaller agencies are disadvantaged because they traditionally do not have the same financial or human resources as larger agencies but are subject to the same environmental and safety regulations.

The information presented in this section is intended to aid in the establishment or enhancement of a pumping station maintenance program. It presents proven concepts and techniques that have been developed over time and can be adopted, improved, or modified for an individual operation, regardless of size or location.

Finally, this section focuses primarily on the more typical types of pumping stations, rather than those serving systems such as low-pressure vacuum or grinder sewers, which are discussed elsewhere (WEF, 2008).

5.1 Increasing Importance of Pumping Station Maintenance

Careful consideration and maintenance of pumping stations have become more important in recent years for several reasons. Regulatory requirements, for example, have become increasingly stringent to reduce or eliminate overflows in the event of station failure. These requirements have led to more reliable and sophisticated station designs, such as portable or onsite standby power systems, duplicate overhead power feeds, and increased equipment redundancy.

In addition, liability for system failures has increased. Large fines have been levied against agencies and individuals responsible for collection systems when, for example, a station failure has resulted in a violation that could be attributed to poor maintenance programs or negligence.

Formation of regional agencies, which is now common, allows more efficient and cost-effective collection and treatment of wastewater. This has resulted in the installation of larger pumping stations and force mains with higher capacities and operating heads. Pumping stations with design capacities of more than 630 L/s (10 000 gpm) and total dynamic heads greater than 60 m (200 ft) are not unusual, nor are force mains longer than 16 km (10 miles). Design techniques and construction methods have changed as a result. Failures (when an overflow or spill occurs) of these larger facilities tend to have a more adverse effect than those of small, older facilities.

Computer-based systems designed for pumping station applications are reliable, cost effective, and functional in many new or rehabilitated pumping stations. Programmable logic controllers (PLC) are used as supervisory pump control systems. Supervisory control and data acquisition systems (SCADA) provide continuous monitoring of the operating status of stations and have the capability to allow station control from a remote location. Instantaneous station alarm reporting, by leased line or radio telemetry systems, allows for faster response times and improved equipment use, and reduces the risk of overflows. The PLC and SCADA systems can be integrated to provide a remote, real-time pumping station control and monitoring system.

Improved design and reliability of variable-speed pump systems provides O & M flexibility and energy savings. Standby or emergency power generators with automatic transfer switches frequently are required during design or retrofitting of pumping stations. These generators represent significant capital expenditures and increased maintenance requirements.

These changes in pumping station design have produced complex systems that have more sophisticated O & M requirements because of their complexity and related environmental and liability considerations. Consequently, those involved with collection systems design, construction, O & M, or management need more and better skills.

5.2 Types of Pumping Station Maintenance

Maintenance programs can take several forms, as outlined in the section below.

Corrective maintenance refers to operation of equipment until it fails and requires repair or replacement. The obvious disadvantage of relying solely on a corrective maintenance program is that it is difficult to schedule maintenance activities or allocate and plan budget and staffing resources. Even more important, the failure of a critical piece of equipment in the pumping station (for example, the failure of a pump bearing from lack of lubrication) could result in overflow or flooded basements upstream. Corrective maintenance based programs are characterized by several features:

- Unscheduled equipment failures,
- Excessive repair or replacement costs,
- Poor use of human resources,
- Lack of control of expenditures,
- Poor morale, and
- Increased liability because of negligence.

Emergency maintenance is a form of corrective maintenance that occurs when a piece of equipment or system fails and creates a threat to public health or equipment. This is a special case of corrective maintenance in which the unscheduled failure of a system or piece of equipment could have a severe or catastrophic effect. Examples of emergencies would be accidental discharges to lakes or streams, backups into homes or businesses, or loss or destruction of equipment.

The cost of emergency maintenance in terms of time and staff resources can be enormous. Furthermore, regulatory agencies frequently view an overflow as a discharge permit violation. Substantial fines and adverse publicity can result. Other potential costs for this type of maintenance are contractor mobilization fees, costs of repair or replacement of destroyed or damaged equipment, cleanup and repair of property where sewer backup has occurred, overtime costs, or costs related to lawsuits.

Preventive maintenance includes routine, scheduled activities performed before equipment failure to extend equipment life, reduce maintenance costs, and increase reliability. Maintenance programs should devote an appropriate level of resources to preventive maintenance.

Experiences in both the public and private sectors show that as more resources are dedicated to preventive maintenance, fewer resources are required for corrective and emergency maintenance, and total maintenance costs begin to decline. Effective preventive maintenance based programs are characterized by:

- Reduced overtime costs,
- Reduced material costs,
- Reduced capital repair or replacement costs,
- Reduced enforcement actions by regulators,
- Improved use of human resources,
- Improved use of financial resources,
- Improved morale, and
- Improved public relations.

When implementing a maintenance program it's important to ask: "at what point does a preventive maintenance program cease to be cost effective?" This question, unfortunately, is difficult if not impossible to answer quantitatively because the effect of a poor preventive maintenance program on emergency maintenance costs is unpredictable. Experience does show, however, that the initial expense that preventive maintenance incurs typically is repaid in longer, more trouble-free operation. The following example may illustrate the benefits of preventive maintenance versus the unpredictable cost of emergency maintenance.

In this example, a failure results in overflow to an environmentally sensitive stream. Emergency expenditures are needed not only to restore operation but also to cover cleanup costs and the resulting fine levied for a permit violation. Bad publicity, an adverse opinion of management of facilities, and budget effects also occur. These costs could easily far exceed the entire annual O & M budget for the station.

Maintenance programs require consideration of other criteria in conjunction with budgetary factors to determine the level of maintenance to be performed. Another possible approach is risk management, or determining what maintenance has to be done to reduce the risk of failure to a reasonable level. Examples of other criteria that typically are considered are equipment reliability, peak-flow capacity, effects of a station failure, frequency and magnitude of failures, and design deficiencies.

Predictive maintenance programs increasingly are used in the public sector. Instruments required for this type of system have become more portable and less expensive. Predictive maintenance is well suited to pumping station maintenance. Predictive maintenance is a form of preventive maintenance that measures the condition of a piece of equipment or machine and then predicts the time remaining before failure occurs. As a result, corrective maintenance can be scheduled on a more rational basis and at less cost. A variety of diagnostic techniques are used to assess equipment. Examples are discussed in the following sections.

Computer-based vibration analysis requires the use of microprocessor based portable field instruments to measure the frequency versus amplitude of equipment vibration. The information can be analyzed onsite and downloaded to a computer to establish and store baseline vibration levels versus frequency. This method allows the detection of equipment problems earlier than is possible with other indicators such as heat or audible noise.

Vibration analysis can detect a variety of electrical and mechanical problems:

- Rotation unbalance;
- Bent shafts;
- Defects in bearings, inner or outer races, and balls or rollers;
- Mechanical loosening; and
- Motor electrical and magnetic anomalies.

Baseline information ideally is obtained as part of the acceptance process for new pumping stations or when the equipment has been overhauled. Periodic measurements compare the vibration levels with the original baseline data and, based on the rate of change, predict failure. Vibration levels can also be compared to industry standards that indicate equipment condition as a function of type or class and operating speed.

Another effective method of predictive maintenance is the use of scanning infrared radiation to detect temperature changes. Small, portable cameras, designed to measure radiation levels and convert these to temperature, are used to scan electrical and mechanical equipment and identify "hot spots". This process is particularly useful for locating potential problems in electrical systems. Infrared scanning typically is used for transformers, main disconnects, branch circuit breakers, motor starters and overload relays, conductors, motors and motor connections, fuses and fuse holders, and capacitors.

Predictive maintenance is a valuable and cost-effective tool that can be applied at stations where reliability is critical and the consequences of an equipment or station failure could be significant. It is not necessary for an agency to buy the diagnostic equipment and train in-house maintenance staff. There are specialized contractors that provide these services.

An effective preventive maintenance program results in the highest reliability and availability of a pumping station. However, even the best preventive maintenance program will not eliminate the need for corrective or emergency maintenance efforts. It is impossible to design a system to account for natural disasters such as hurricanes, earthquakes, torrential rainfalls, or other emergencies. Thus, a good overall maintenance program encompasses elements of all three types of programs.

6.0 PUMPING STATION OPERATIONS AND MAINTENANCE STAFFING

The organization and staffing required to support a pumping station maintenance program depend on electrical and electronic systems, mechanical systems, and hydraulic and pneumatic systems.

Operations and maintenance for each of these systems requires a different set of skills and varying levels of knowledge. Because it would be extremely unlikely for any single employee to have the detailed knowledge required to operate and maintain all of these systems, it is typical for an agency to separate O & M employee classifications.

Operating responsibility, which would encompass routine maintenance of the pumping station, is assigned to one group of employees. System and equipment maintenance is assigned to a second group, whose members have specialized skills in electronic, mechanical, hydraulic, and pneumatic systems. The second group might include pipe fitters, electricians, and machinists. Typically, the tasks to be performed and the skills required to perform the tasks form the basis of the classifications. It is not necessary to have a separate classification for each craft if the organizational structure addresses what work is to be done, what skills are necessary to do the work, and who is going to do it.

Many state and local codes require formal education, certification, or licensing to work on specific systems. Because of safety issues, maintenance of electrical, low- and high-pressure hot water, or steam boiler systems may require a state or local license.

Optimal staging of staff, equipment, and space requirements at different service areas will reduce fuel consumption, toll costs, response times, and travel distances. Staging presents an opportunity for a smaller work force, more available work functions in each service area, and an increase in wrench time. Through staging, the staff will become more intimately familiar with their smaller service areas. The increase in communication between the pipeline and pumping personnel in each service area will improve the effectiveness of management. Staging often results in the benefits of cross-training and team-building.

Retention of system knowledge has become an increasingly significant discussion item as key employees near retirement age. One method for retaining knowledge is to train individuals and increase their knowledge base to slow down the "brain drain" of retirement. The training sessions should be offered for various levels of responsibility or functions; many training sessions should revolve around everyday collections and pumping functions. A catalogue of classes should be kept online for all employees to review and to serve as a reminder of what is available. The classes should be include a title, objective, and targeted attendee and should cover:

- Overview of pumping and collections groups,
- Generators,
- Hand, and power tools,
- Mapping and GIS,
- Basic computer skills,
- Towing and tie-down of equipment,
- Reports and forms,
- Effective meetings,
- Construction practices,
- Televising,
- Odor control, and
- Wet-tapping and line stops (plugging and/or bypass pumping).

6.1 Pumping Station Operations Staffing

Typical pumping station operator responsibilities may include:

- Performing routine pumping station inspections;
- Recording or interpreting information from elapsed time meters;
- Alternating pumps;
- Performing station housekeeping;
- Cleaning bar screens and wet wells;
- Verifying flow information;
- Exercising emergency standby power systems;
- Adjusting packing glands;
- Testing alarm systems;
- Checking parameters such as incoming voltage and current using hardwired meters;
- Performing equipment preventive maintenance such as lubrication, belt tightening, checking of fluid levels, and exercising of equipment on a routine schedule;
- Responding to alarm conditions at the pumping station;
- Diagnosing problems and requesting assistance to resolve them when necessary;
- Initiating work orders;
- Assisting craft maintenance crews;
- Providing input on design or station rehabilitation;
- Participating in the budgeting process to identify resources; and
- Supervising field operations.

There may be from one to four levels of experience in the operator classification—entry level operator, journeyman operator, line supervisor/lead operator, and manager—depending on the number and size of stations in the system.

Additional responsibilities for these classes frequently include pipeline inspection, cleaning, and maintenance.

Each of these classes has specific job descriptions that include the tasks involved, skills and level of education required, and performance criteria. In establishing a program, many agencies have used certification criteria to define job descriptions.

Certification information is available from many sources, including the Association of Boards of Certification program guides and state certification agencies. Before certification can be used as an incentive for job advancement, it must be strongly supported and encouraged, and sufficient financial resources must be provided.

6.2 Pumping Station Maintenance Staffing

A typical staffing organization includes classes of personnel for craft or trade maintenance to support equipment maintenance of electrical, mechanical, and hydraulic/pneumatic systems including a manager, lead person/line supervisor, and journeyman. Depending on the size, structure, and organization of the agency, the managers for each of the two levels of O & M can be individual managers, one assigned to maintenance and the other assigned to operation, or a single manager responsible for both O & M. A number of agencies also use a fourth class, apprentice journeyman, administered through an established formal apprenticeship program.

Typically, craft personnel are responsible for performing the following:

- Low- and medium-voltage motor control center maintenance,
- Supervisory systems maintenance,
- Instrument and control system maintenance,
- Internal motor maintenance,
- Lighting panel and branch circuit maintenance,
- Pump overhaul,
- Mechanical seal replacement,
- Vibration measurement analysis,
- Pneumatic systems maintenance,
- Valve maintenance and overhaul,
- Engine repair and overhaul,
- Generator maintenance, and
- Heating, ventilation, and air conditioning system maintenance.

Although each member of the O & M crew has specific task assignments based on skills and jurisdictional issues, a successful maintenance program integrates these skills to create a team. For example, the removal of a pump from a pumping station for overhaul may require an electrician to disconnect motor connections, an operator to reset the supervisory control for pump sequencing, and a mechanic for rigging and removing the pump.

Approaching pumping station maintenance as a team assignment facilitates efficient resource use. One effective approach is to develop composite crews on a skills-required basis and use crew members within their jurisdictions. In the preceding example, the electrician and mechanic would be an established crew, assigned work that requires both sets of skills.

Cross-training is another way to increase maintenance effectiveness. Maintenance personnel are often unable to properly diagnose the cause of a pumping station system failure because pumping station systems tend to be specialized. Premature mechanical seal failure in a pump could be the result of excessive vibration in the motor, pump, or drive shaft or a critical frequency in the entire system. Nonspecialized maintenance personnel may have difficulty identifying the source of the problem or may take an inordinate amount of time to do so.

7.0 OPERATIONS AND MAINTENANCE FACILITIES AND EQUIPMENT

A properly planned and supported equipment yard is essential to collection systems operation. It is the location from which equipment, materials, and personnel are dispatched and where operations records are kept. Ideally, these facilities are self-sufficient and independent, except where a central repair yard or heavy-duty repair shop is available.

In smaller municipalities, collection systems maintenance equipment and personnel typically share one yard with other municipal operations, such as water and street departments. In this case, it is important that a portion of the yard be designated for collection systems maintenance activities alone. Specialized tools can be segregated or color coded so that they are not used on water supply systems. This reduces confusion regarding equipment use and repair. All essential tools and portable equipment should be kept in a secure location, with keys issued to supervisory personnel.

The area should be planned carefully; often the yard is small, and the opportunity for confusion increases as the space diminishes. If facilities are properly placed, interference between departments is reduced. The layout and location of the yard are reviewed based on current use and traffic conditions and current and future land values.

Satellite yards to house specialized collection systems equipment can reduce response time on weekends or holidays. Proper response to collection systems complaints will improve public relations; the primary goal is to restore service promptly in the event of interruptions. In large cities, considerable time can be lost because of traffic delays or distances between the yard and job site. Agencies in growing metropolitan areas typically prepare a long-range program and divide the area into districts, each with a central yard.

Yard design should provide the proper housing of equipment and materials affected by inclement weather. Special storage space can be allotted for concrete, bricks, and other small materials. Raised platforms inside and outside accommodate the on- and off-loading of daily equipment, materials, and vendor deliveries. It is recommended that standard or special platforms match the truck-bed level. A platform-mounted, hand-operated hoist is useful for heavy lifting.

Pumps, compressors, mixers, generators, bypass piping, and other items are best stored separately so that available equipment is visible at a glance. Emergency equipment used at irregular hours (lights, bypass pumps, etc.) typically are kept away from regularly used tools so that they can be found quickly when needed.

Storage bins for sand, gravel, dirt, and asphalt should be wide enough to fully accommodate the bucket of a front-end loader or backhoe so that the material may be quickly handled and moved. Bins typically are deep enough to hold at least a one-month supply of materials if used regularly or to hold enough material to complete two repairs if used only for that purpose. Materials stored outside may be damaged by the elements over time and can be washed away during extreme wet-weather events. A cover for these materials may be needed to reduce weather-related damage, especially if the area has a history of extreme wet-weather conditions.

A clean, orderly yard can strengthen morale and efficiency. Time-controlled lighting for safety, security, or night operations is essential. Equipment should not be stored dirty or in disrepair. A wash rack typically is provided for cleaning trucks and equipment. It is recommended that equipment, vehicles, and storage areas be subject to routine unannounced inspections. An internal award system can establish a competitive spirit between departments that share a common yard.

The office, which may or may not be connected to other buildings, typically is large enough for the proper storage of maps; survey charts; reports and records; and office materials such as cabinets, computers, printers, reference materials, communication equipment, and furniture for staff and visitors. There can also be room for small, specialized tools or computer systems. It is important that items such as transits, laser lights, levels, and footage counters be properly secured at all times when not in use.

Although agencies may still want to maintain public records at the yard, computer systems increasingly are being linked with these records for immediate recall to an administrative facility offsite. To promote the best image and preclude disruption of yard activities, daily public contact will most likely take place at the administrative facility rather than at the yard office, although this depends on system size. Growth may play a role in planning a new facility or reexamining an existing one. For example, if the agency grows larger, more space will be needed for materials, equipment, and personnel. If the facility is no longer large enough to accommodate needs, a satellite yard can be established.

Interior walls, floors, tables, chairs, and counters are constructed or treated with materials that resist wear and can easily be cleaned. A separate room for removal and cleaning of rain or snow gear and boots will help lengthen the life of this equipment and keep the rest of the facility clean. It is important to keep in mind that the administrative offices, whether in a central or satellite facility, are open to the public. A clean and orderly appearance can be the best public relations image an agency can present.

Central repair maintenance shops vary in size and scope. Some agencies maintain the sanitary collection system only, whereas others are responsible for storm sewers, pumping stations, treatment plants, or street repairs. In smaller communities where street, water, and collection systems departments are consolidated in public works, it may be more economical to have a central repair shop. At the common yard, all equipment maintenance would be performed and all departments would operate. A utility truck can be set up with an assortment of tools; parts; and welding, cutting, and hoisting equipment so that the mechanic can go to the problem when necessary.

8.0 REFERENCES

Water Environment Federation (2008) *Alternative Sewer Systems,* 2nd ed.; Manual of Practice No. FD-12; McGraw-Hill: New York.

Water Environment Federation; American Society of Civil Engineers (2008) *Existing Sewer Evaluation and Rehabilitation,* 3rd ed.; Manual of Practice No. FD-6; ASCE Manuals and Reports on Engineering Practice No. 62; McGraw-Hill: New York.

Chapter 3

Information Management

1.0	INTRODUCTION	48	2.7	Preserving the Utility's Collection System Corporate Memory	53
2.0	ESTABLISHING AN INFORMATION MANAGEMENT SYSTEM	48	3.0	BUILDING AN INFORMATION MANAGEMENT SYSTEM	53
	2.1 Conformance to Regulatory Responsibilities	49		3.1 Types of Collection System Information	53
	2.2 Performing Timely and Consistent Information Searches	50		3.2 Selecting System Attributes to Retain in a Collection System Information Management System	54
	2.3 Maintaining Perpetual Calendars for Preventive Maintenance and Inspections	51		3.3 Practical Guidelines for Establishing an Information Management System	57
	2.4 Providing Justification for Operations Budgets	51	4.0	OPERATING AND MAINTAINING AN INFORMATION MANAGEMENT SYSTEM	58
	2.5 Tracking Prioritized Work Order and Repair Schedules	52			
	2.6 Organizing Capital Rehabilitation and Replacement Plans Based on Asset-Management Data	52	5.0	REFERENCE	58

1.0 INTRODUCTION

This chapter focuses on the information management needs and requirements of the collection system manager. Presented here are general management principles-rather than complex technical elements—based on the assumption that technical and supervisory resources are available elsewhere to implement the necessary software and hardware solutions.

Three forces have arisen that are compelling collection system managers to evolve away from paper-based (or non-existent) recordkeeping to computer-based systems: federal and state environmental regulators, such as U.S. Environmental Protection Agency (U.S. EPA); asset-management requirements (Government Accounting Standards Board [GASB] 34); and overall improvements to the industry (technical certification programs). In theory, paper-based systems can fulfill the analysis and reporting requirements of these three demand areas, but the cost of manually collecting and reporting the data is not financially acceptable in light of the availability of contemporary computer-based systems.

However, the evolution towards computerization has not been without its problems. For example, computerization can create a flood of raw data from multiple sources that can result in an incoherent picture, thus becoming a source of frustration and difficulty—rather than assistance—for a collection system manager. However, when properly filtered and sorted, the data can become *knowledge,* which when combined with a collection system manager's experience, can yield *insight,* leading to effective resource decisions. The objective of this chapter is to provide the reader with a sufficient framework with which to make those decisions in a balanced and professional manner.

2.0 ESTABLISHING AN INFORMATION MANAGEMENT SYSTEM

Collection systems have historically been viewed by the public and elected officials as low-tech and somewhat simple in nature. Therefore, establishing a high-tech electronic information management system (IMS) can be more challenging for the collection system manager to execute than say, for example, the purchase of a sewer cleaning machine, which is a much more routine and familiar proposition. An IMS requires a noticeable investment to establish and maintain, including the cost of staff labor, and to a lesser degree, software and hardware. Such an investment will be a new, added, and permanent cost for the utility, and these factors tend to draw the scrutiny of elected governing boards.

To make the establishment of an IMS as palatable as possible to governing boards, the manager needs to develop a case based on the highest possible goals and rationales rather than on just the technical merits. Emphasizing the overall public benefit of an IMS will engender the broadest political and community support, compared with the more limited success that might be achieved if the merits are described using complex, difficult-to-understand computer-based terminology.

To assist the manager in developing a rationale for a governing board, the following are the most common and defensible principles upon which to establish a collection system IMS:

- Conform to regulatory responsibilities, principally U.S. EPA's capacity, management, operations, and maintenance (CMOM) program and its variants;
- Perform timely and consistent information searches;
- Maintain perpetual calendars for preventive maintenance and inspections;
- Provide justification for operations budgets;
- Track prioritized work order and repair schedules;
- Organize capital rehabilitation and replacement plans based on asset-management data;
- Preserve the utility's corporate memory;
- Provide real-time, user-friendly information to field staff; and
- To provide useful and timely maintenance trend and productivity reports.

2.1 Conformance to Regulatory Responsibilities

Since the mid-1990s, collection system agencies have been subject to varying degrees of regulatory requirements embodied in CMOM. An IMS is an integral part of CMOM because of the accuracy and consistency that can be achieved with well-structured and executed computerized management programs.

CMOM programs, as described in the U.S. EPA *Guide for Evaluating Capacity, Management, Operation, and Maintenance (CMOM) Programs at Sanitary Sewer Collection Systems,* incorporate many of the standard operation and maintenance activities that are routinely implemented by utilities (U.S. EPA, 2005). It does, however, create a new set of information management requirements to:

- Better manage, operate, and maintain collection systems;
- Investigate capacity constrained areas of the collection system;

- Proactively prevent sanitary sewer overflows (SSOs); and
- Respond to SSO events.

The IMS requirement has been incorporated by numerous state and regional regulatory bodies into their own versions of CMOM, and these requirement(s) can be relied upon to form the development and use of an IMS within the collection system utility.

2.2 Performing Timely and Consistent Information Searches

The speed with which information can be retrieved from the Internet has raised the public's and media's expectations how local government delivers data. This has obvious implications for collection system managers who are subject to the same expectations for speed. It makes the processing power of an IMS an invaluable asset that collection system agencies cannot do without because poring over reams of paper records to retrieve, collate, and distribute information is a grinding, error-laden task that does not speak to the highest possible professionalism of a collection system utility. Outlined below are the most frequent demands for information retrieval.

- Customer and media inquiries are almost always related to complaints or problems because there is almost no interest on the part of the public in the normal operations of a collection system. The ability to provide an immediate reply using information retrieved electronically leads to high marks awarded for customer service.

- Elected officials expect to be kept informed on the performance of the collection system and its engineering and operations staff, especially during rate-increase requests. A well-organized database can yield compact, uniform reports that bolster the confidence of elected officials and provide a sound basis for making resource decisions.

- The Government Accounting Standards Board (GASB) promulgated Rule 34 to measure the degree to which the infrastructure of the United States is being maintained. It seeks to establish a profile of the life span of the nation's infrastructure and when it will need to be replaced or rehabilitated. Although GASB is not an infrastructure management organization, it is responsible for setting the standards by which the financial health of the cities and agencies is measured because they must be able to establish the revenue needed to keep their infrastructures intact and functioning. An IMS designed to yield current and projected infrastructure data can be used to satisfy GASB 34 reporting requirements. Engineering data, asset-management data, inspection records, and maintenance histories form the foundation of GASB 34 reporting.

- When a collection system utility is faced with a damage or liability claim, gathering and arranging historical data is difficult without an IMS. Attorneys who represent agencies in litigation are at a distinct disadvantage when records and reports are not digitally organized. In addition, a utility can depend on its IMS to alert them to repetitive conditions such as flooding patterns that can leave them exposed to damage claims.

2.3 Maintaining Perpetual Calendars for Preventive Maintenance and Inspections

The repetitive nature of preventive maintenance and inspections can create a mind-numbing experience when trying to organize, assign, and report activities without an IMS. Using an IMS to keep track of commitments made over long periods of time helps to ensure reliability and continuity. In fact, preventive maintenance and inspections that have long intervals between activities are the ones most likely to be forgotten without the infinite calendars built into an IMS. Paper reminders or other manual scheduling methods are more likely than not to become lost or forgotten, especially for long-interval work.

The GASB 34 standard has specific timetables for inspection intervals, which can be loaded into the IMS. Because pipe inspections tend to occur at long intervals, an IMS becomes invaluable for this task.

2.4 Providing Justification for Operations Budgets

Operations budgets are typically calculated on the basis of resources needed for each of four areas: preventive maintenance, inspections, complaint servicing, and spot repairs. Without an IMS, many collection system managers are left to define their budgets based on historical experience or intuition, neither of which has solid professional footing.

An IMS that is properly loaded with data can easily create a profile of the correct budget by examining resource requirements that are defined for each of the four areas of operations. For example, a manager can use the IMS to quickly and accurately sum the total labor hours required for a year's worth of preventive maintenance to determine if the budget is sufficient to accomplish all the tasks. If the manager determines that there are 10,000 labor hours of preventive maintenance work to do in a year, but has a budget for only 8000 labor hours, it is easy to see that the utility has an accumulating deficit of work and they will never "catch up". These types of calculations are readily available in an IMS and provide proof-positive to elected officials the need for an adequate foundational budget.

Additionally, an agency that can never keep up with its preventive maintenance schedule could face negative consequences from regulators who would be intolerant of an inherently deficient preventive maintenance program.

2.5 Tracking Prioritized Work Order and Repair Schedules

Prioritization of work orders is a dynamic process because collection systems themselves are hydraulically and structurally dynamic. In addition, environmental factors such as wet weather add to the process. Because an IMS operates in real time, the priority shifts that will occur also can be reported in real time. This provides an accurate picture of the demands on a utility at any given moment. This is especially important during emergencies when a manager will be called upon to explain the service status of the organization. An IMS-based report quickly will be able to display both the specifics and the global status of the situation.

Additionally, linking the status reports of an IMS to the Internet provides the broadest information distribution to the public. This goes far in reassuring the public that the service they are expecting is not lost in an indeterminate "black hole" of local government.

2.6 Organizing Capital Rehabilitation and Replacement Plans Based on Asset -Management Data

Prioritized capital replacement and rehabilitation plans that are linked to asset-management information, are similar to prioritized work order and repair schedules as described above. However the sheer scale and expense of doing capital work requires a separate prioritization scheme that is much more complex than the relatively simple variables associated with work orders.

Several collection system attributes can cause asset management to be complex: structural materials and performance; maintenance and repair history; anticipated future flow; the ability of the agency to finance capital work; and much more. Capturing this information and producing a rational capital management plan based on this degree of complexity is only achievable with an IMS as the cost of assembling and analyzing the information manually is cost prohibitive.

The rise of sophisticated asset management as a refined science is due primarily to the availability of data analysis capabilities associated with computer systems. Until recently, the complexity of this information has been too overwhelming to cost-effectively analyze using other means.

2.7 Preserving the Utility's Collection System Corporate Memory

Of all the principles upon which to establish and maintain a collection system IMS, preservation of corporate memory is perhaps the most important of all. A collection system is a long-lived public asset and the knowledge of how the system was built, operated, maintained, and rehabilitated/replaced is a knowledgebase that will outlive current and future utility staff.

Too often, the retirement of a knowledgeable employee has meant that a substantial part of the utility's corporate memory was lost, to the detriment of those still working there and to the ratepayers who support the system. For this reason alone, an IMS that can maintain this knowledge across generations of employees is a valuable asset for a manager.

3.0 BUILDING AN INFORMATION MANAGEMENT SYSTEM

Once the premise and rationale for an IMS has been established and approved, the process of building it can begin. This process involves an understanding of the types of data associated with collection systems, the attributes that need to be captured, and the reports that are generated.

3.1 Types of Collection System Information

Building IMS for a collection system involves an understanding of the different types of data found in a typical system. In analyzing the information management needs of a utility, the manager should be familiar with the basic types of data that are typically associated with collection systems.

Collection system information can be categorized into five principal types:

(1) *System data* is the heart of collection system information, upon which all other data are dependent. System attributes consist of an inventory of a collection system's assets, its maintenance, repair, inspection, and condition assessment. Customer service records and the history of complaint

resolution are also components of system data. (Failure to maintain customer service data may invite litigation as a result of claims for damages.)

(2) *Spatial data* describes the location of system structures with respect to real world geographical references, typically survey coordinates, and often are derived from global positioning satellite readings. Spatial data is the backbone of geographic information systems (GIS), which are used to create electronic maps of collection systems and link data to them.

(3) *Dynamic* data is the reporting of real-time data that describes how the collection system is functioning. The most common dynamic data is pumping station activity and alarms reported by supervisory control and data acquisition systems; flow metering results; and rain gauge readings.

(4) *Modeling* data is used to determine the existing and future hydraulic capacity of a collection system. It is derived from sophisticated computer software that melds inventory parameters, such as pipe size, with topographic data, such as slope, with flow metering results and storm activity to profile the system's ability to handle flow during dry weather and wet weather events.

(5) *Asset value* data is the calculation of the monetary value of the collection system, the cost to rehabilitate or replace it over its lifetime, and the revenue demands necessary to support those capital improvements. There is a close link between how the system is maintained and the length of its life expectancy.

3.2 Selecting the System Attributes to Retain in a Collection System Information Management System

A collection system manager should aim to have a set of criteria by which to judge whether or not the utility should expend resources to capture, store, and report an attribute. The goal should be to gather a sufficient, but not overwhelming, amount of information to allow a reliable understanding of the collection system. The five basic collection system attribute groups are listed at the end of this section.

It is important that before the information system is built, the manager defines the ongoing daily, weekly, monthly, and yearly reports that will be required from the IMS. These output reports will dictate the input requirements of the IMS, which will eliminate gathering and inputting information of questionable value. The manager should start with the *end* in mind, because waiting until the system has been implemented to try and design ongoing reports will be expensive to correct.

Also, there must be an appropriate balance between the capability of the software and the capability of the agency using it. The sophistication of the desired software should not exceed the staff's ability to use it, as this could result in a significant loss of investment. Conversely, agencies need to closely analyze the promised capability of proposed software installations to ensure that it will actually deliver the desired functions.

To help facilitate the attribute selection process, a collection system manager should ask several questions about each. For example, does it identify a unique value, such as a structure's identification or location, or a customer? Unique values are the most important attributes to capture because databases and maps depend on this data to navigate finding, sorting, and filtering data. Or is it reportable information that can be counted, summed, or averaged?

Statistical analysis and reporting are the primary products of an IMS because they exploit the calculating speed of computers. Incidental information that would not be subject to statistical analysis may not be worth the effort. For example, is it worth the effort to know the size of each manhole cover in a system?

There are several good examples of statistical data worth keeping and reporting:

- A daily list of incomplete work orders,
- A sum of the length of all the pipe in the system,
- The number of SSOs per approximately 160 km (100 miles) of pipe in the system, and
- The number of service laterals in the system.

Other questions include: Is the attribute a problem flag? Does the attribute serve to alert you to an important condition that is otherwise hard to remember? Does it warn you of an effect or potential effect on public health and safety, or on the safety of crews?

For example,

- Does a structure have an exceptionally hazardous aspect, such as a nearby industrial waste discharger?
- Is it a work-order location that is known to be barricaded and, therefore, more hazardous to the public?

Does it inform the manager of important contractual or ownership issues regarding a sewer structure?

For example,

- Is it a structure that is privately owned or maintained, and is, therefore, not the utility's responsibility?
- Is a structure's construction guarantee still in effect when a repair is required?

Is the attribute map-dependent data, and is needed for building and maintaining the collection system maps?

Examples of map data are:

- Latitude/longitude survey data,
- Rim elevation,
- Invert elevation,
- Depth, and
- Length.

If the attribute is an inspection defect, then is the defect significant enough to be worth keeping in the IMS?

Inspection defects are the single most abundant attribute in a collection system IMS. They are usually derived from television inspection records, and with inspection equipment it is possible to capture endless amounts of defect observations.

However, it is not possible or realistically desirable to repair every observed defect because the cost would not be justifiable to perform trivial repairs. Collection systems, like all infrastructures, are in a constant state of change and the most cost effective maintenance and repair is that which addresses significant problems, not all problems.

Therefore, it would likely be in a collection system's best interest if only the significant defects were entered into the IMS to prevent the database from being polluted by trivial, distracting, or non-consequential information.

There are five basic collection system attribute groups that would comprise both an IMS and a GIS:

(1) Structure data.
   - Manholes,
   - Main lines,
   - Service laterals,
   - Catchbasins or stormwater inlets,
   - Pumping stations and comminutors,
   - Septic tanks and miscellaneous structures, and
   - Stormwater storage/detention facilities.

(2) Maintenance data.
   - Scheduled preventive maintenance,
   - Scheduled repair work orders,

- Unscheduled/emergency work orders, and
- Weekly, monthly, and yearly maintenance and crew productivity reports.

(3) Inspection data.
- Structural defects
  - Pipeline defects—cracking, deflections, etc.,
  - Pipe joint defects—offsets, etc., and
  - Manhole and catchbasin/stormwater inlet defects—structural components.
- Maintenance conditions
  - Debris,
  - Grease,
  - Roots,
  - Odor, and
  - Vermin.
- Inflow and infiltration measurement.

(4) Hydraulic/capacity defects—surcharging; only a subset of the factors needed to determine optimum capacity.

(5) Condition assessment—a combination of specific physical conditions such as defects and structural data such as material type.

3.3 Practical Guidelines for Establishing an Information Management System

One of the most important mantras to remember is: garbage in, garbage out. Installing a computer system cannot repair the errors of a poorly managed, paper-based system. If a utility's paper records are in disarray, mere data-entry of those records will only exacerbate the problem of trying to remedy the errors. Like house painting, the majority of the work is in the preparation, not the final step.

The biggest cost of computerization is data collection, input, and upkeep—not software and hardware. When a utility undertakes the establishment (and operation) of an IMS, there is a tendency to underestimate the labor hours required to build and use the system, including staff training. Those labor hours, which are a real cost, cannot by trivialized and should be a primary focus when considering an IMS. The cost of the IMS software and hardware, even some of the most expensive systems, does not even compare to the required labor commitment.

4.0 OPERATING AND MAINTAINING AN INFORMATION MANAGEMENT SYSTEM

Operating and maintaining the IMS is vendor-specific and should be addressed by its supplier. However, there is one aspect of the IMS that only the collection system manager can supply: security of the data against accidental loss because of fire, employee mistakes, or vandalism.

Every labor hour that is put into building and using the IMS increases the economic worth of the system in replacement-value terms; the IMS could eventually become the utility's single most valuable asset.

Therefore, the collection system manager must use his or her authority to provide safe, off-site storage for duplicate copies of the database(s). Merely keeping duplicates in a file cabinet next to the computer will be useless in the event of a catastrophic fire in an office building. And even if a recovery were attempted from such a loss, anecdotal attributes could never be recovered.

5.0 REFERENCE

U.S. Environmental Protection Agency (2005) *Guide for Evaluating Capacity, Management, Operation, and Maintenance (CMOM) Programs at Sanitary Sewer Collection Systems*; EPA 305-B-05-002; U.S. Environmental Protection Agency, Office of Enforcement and Compliance Assurance: Washington D.C.

Chapter 4

Collection System Assessment and Capital Improvement Planning

1.0	INTRODUCTION	60
2.0	SYSTEM ANALYSIS	61
	2.1 Capacity Assurance Planning	61
	2.2 System Analysis Tools and Methods	62
	2.2.1 Data Collection and Management Methods	62
	2.2.2 Supervisory Control and Data Acquisition Systems	63
	2.2.3 Geographic Information Systems and Computer-Assisted Drawing	63
	2.2.4 Computerized Maintenance Management Systems	64
	2.2.4.1 Maintenance History	64
	2.2.4.2 Complaints and Service Requests	65
	2.2.4.3 Work Orders	66
	2.2.5 Integrating Geographic Information Systems and Computerized Maintenance Management Systems	66
	2.2.6 Inflow and Infiltration Identification	67
	2.2.7 Flow Monitoring/ Monitoring	69
	2.2.8 Hydraulic and Hydrologic Modeling	72
	2.2.8.1 Steady-State Simulation versus Dynamic Simulation	73
	2.2.8.2 Flow Development	74
	2.2.8.3 Calibration/ Verification	75
	2.3 Condition Assessment	76
	2.3.1 Surface Inspection	77

(continued)

	2.3.2	Closed-Circuit Television	78		3.2.1.2 Performance 88
					3.2.1.3 Risk 88
	2.3.3	Smoke Testing	79		3.2.2 Validation 90
	2.3.4	Trunkline Inspection	80		3.2.3 Replacement/ Rehabilitation Goals 91
	2.3.5	Lift Stations	81		
3.0	CAPITAL PLANNING		82		3.2.4 Industry Benchmarks 91
	3.1	Master Planning	84		3.2.5 Prioritization 92
	3.2	Capital Improvement Plan Development	85		3.2.6 Final Product 93
				4.0	REFERENCES 94
		3.2.1 Ratings	85		
		3.2.1.1 Condition	88	5.0	SUGGESTED READINGS 94

1.0 INTRODUCTION

A well-documented capital improvement plan (CIP) provides a utility with a sound understanding of the costs and schedules of system improvements that will be required to sustain performance in the future. An accurate CIP allows managers and decision makers to understand near and long-term financial needs, which is critical to sustaining a consistent level of service at a consistent cost to customers.

This chapter will outline the process of developing and updating a CIP and describe the information required to conduct this process. Because CIP development requires an understanding of both existing system conditions and projected future needs, this chapter includes a discussion of system assessment approaches and technologies. Capital improvement plans are a multi-year planning instrument used to identify needed capital projects and to coordinate funding needs and timing to maximize value to the public. The key components to a successful CIP include:

- Reliable tools for inspection, monitoring, and recording the condition of the collection system infrastructure;
- Methods and tools to evaluate capacity and the effects of future growth on the system;

Collection System Assessment and Capital Improvement Planning

- System to incorporate and prioritize input from the key stakeholders interests including engineering, operations, finance, management, and customer;
- Priority matrix developed with input from the key stakeholders; and
- Appropriate funding levels and phasing of improvement strategies to achieve desired replacement goals.

This chapter guides the wastewater collection system manager through the important steps in developing a CIP. The CIP process includes developing funding requirements that will be used for budgeting, which is discussed in more detail in Chapter 8. In general, to develop a CIP a manager needs to follow several key steps:

(1) Identification,
(2) Validation,
(3) Prioritization, and
(4) Financing (see Chapter 8).

2.0 SYSTEM ANALYSIS

Understanding the system is the first step in developing a successful CIP. Various tools and methods are available to help the manager gauge the pulse of the collections system. These tools are vital in identification and prioritization of system improvements.

2.1 Capacity Assurance Planning

Most municipalities have been experiencing population growth and, as a result, an increase in flow to existing infrastructure. In addition, infiltration and inflow (I/I) increases as infrastructure ages; these increased flows can result in basement backups and sanitary sewer overflows. Hydraulic models and use of other analysis tools—including spreadsheets and Manning's equation—typically are used to identify size alternatives to relieve capacity restrictions for various storm and buildout conditions. Other tools such as geographic information systems (GIS), flow monitoring and supervisory control and data acquisition (SCADA) are useful tools in identifying potential capacity issues with an increase in growth and I/I.

Managers also should track incremental flows added to the system, which can affect capacity. The capacity of all new developments should be recorded to eliminate the possibility of "double dipping" by allocating the same capacity to two different developers based on the timing of actual construction. This can be accomplished using GIS or a simple spreadsheet as new connection fees are collected. The data can then be used for various modeling simulations to determine the potential effect on the system.

2.2 System Analysis Tools and Methods

The process of developing a CIP has evolved over the years as new tools and methods for assessing wastewater collection systems have emerged. Tools now used for development of a successful CIP include hydraulic models, GIS, automated field data collection, asset-management systems, and sophisticated databases. These tools, when used together, can assist the manager in making well-informed, defendable, and optimized decisions about which projects to include and what priority each receives.

2.2.1 Data Collection and Management Methods

Acquisition, analysis, and application of collection systems information is the foundation of accurate and effective decision-making in a CIP. The most effective way to collect and analyze this information is by using an asset-management system. A utility's ability to effectively manage its infrastructure is directly linked to its ability to maintain accurate and up-to-date information on the status and location of its facilities. At a minimum, this requires that changes made by field crews, such as repairs and small replacements performed through operations and maintenance (O & M), be recorded and shared with staff maintaining the facility's maps, records, and engineering information.

Information about a collection systems inventory should be linked to its geographic location to provide a spatial context of the extent and condition of the system. This link between the data and spatial component is achieved through the use of GIS. Through the use of unique identification numbers, geocoded addresses, and global positioning system records, assets can be viewed in the correct geospatial context for a variety of uses, including locating where critical assets exist and how they are grouped.

Asset-management systems and GIS are useful tools for developing a CIP only if the data that populates these systems is accurate, reliable, and reproducible. Various data collection methods are available to achieve this goal including SCADA, global position systems (GPS), field data collectors, flow monitoring, and collection system inspection techniques (closed-caption television [CCTV], smoke testing, manhole inspections, sonar inspections).

2.2.2 *Supervisory Control and Data Acquisition Systems*

Supervisory control and data acquisition systems provide continuous monitoring and control of remote points. In wastewater, SCADA typically is used to monitor collection system lift stations and to provide real-time alarming and historical archive of system operations for analysis and predictive maintenance. Data collected through a SCADA system provides key information for development of a CIP, including pump run times, increase in head pressure or volume of I/I flow, and a history of system deficiencies identified by real-time alarming. The SCADA helps the wastewater collection system manager identify future problems and plan and budget for future CIP projects.

2.2.3 *Geographic Information Systems and Computer-Assisted Drawing*

One of the most common problems in collection systems management is determining the locations of collection system lines and manholes. This task is best done by keeping maps showing the locations of the collection system lines, cleanouts, and manholes; depth-to-sewer invert; pipe material and size; and the direction of flow. Pumping stations, appurtenances, and wastewater treatment plants (WWTPs) should be clearly identified. With the popularity of mobile computers, maps of the collection system can be made available to field personnel. Maps also can show the location of other utilities like water power and natural gas.

System mapping for many utilities consists of large, paper maps divided into overlapping, large-scale sections. These maps or sheets are bound into books that can be stored easily and taken into the field as needed. For older systems, original maps may not be accurate because construction over the years may have relocated entire sections. If not maintained properly, maps can quickly be outdated by development, reducing accuracy and confidence-level.

Developments in computer mapping have made it possible to detail systems and easily add new structures. Computer-assisted drawing and GIS applications have expanded to offer a wider variety of uses, from simple mapping to computer-aided modeling for predictive management. Among some of these parameters exist the ability to provide:

- Information layering and dimensional,
- Variety of maps for different collection system applications or designs,
- Easily revised maps,
- Record of activity sites and causes of problems,

- Record of current and future rehabilitation sites,
- Record of current and projected maintenance sites, and
- Record of television inspection sites.

2.2.4 *Computerized Maintenance Management Systems*

Collection systems contain several assets that need to be managed and maintained: sewer mains, manholes, siphons, forcemains, cleanouts, lift stations, and laterals. Some regulatory bodies that manage discharge permits are requiring agencies to have a system in place to schedule regular maintenance and cleaning of the sanitary sewer system and document these activities. As discussed in earlier chapters, asset management is a process that addresses the efficient, long-term management of the collection system from an engineering and economic perspective. Asset management is a sustained, systematic process of designing, operating, and maintaining assets and facilities effectively. It is important to remember that asset management is not a software system but rather a programmatic approach to managing and maintaining assets that is accepted and supported by the entire organization. Computerized maintenance management systems (CMMS) are software packages that can assist an agency in maintaining a computer database of information about an organization's maintenance operations. This information is intended to help maintenance workers do their jobs more effectively and to inform management decisions. Today, most systems are computer-based and easy to use. Advances in computer technology have made the development and maintenance of databases cost effective for even the smallest utilities. Systems should be tailored to meet unique facility needs. A careful evaluation of specific needs, based on clear objectives, should development of an effective system.

Specific categories of information necessary for computerized maintenance systems are facilities inventory, service work order, and maintenance history. Inventory data should be available within the CMMS so that jobs can easily be planned and crews can be sent to locations with the appropriate parts and materials. This information is important to the organization for other uses, such as asset management, capital planning, or hydraulic modeling. Historically, location and attribute data have been maintained on maps and drawings, inventory cards, or manual records. The data conversion activity will be the largest effort required of most utilities.

2.2.4.1 *Maintenance History*

Maintenance history data are valuable for determining budgets, planning capital improvements, and developing preventive and predictive maintenance activities.

Maintaining records serve two purposes: to help record work done in the system and to help plan future projects. Records of the maintenance activities are typically tracked on prepared forms corresponding with a particular CMMS format. These forms include information such as the date, location, work performed, distance, reason for work, material removed, equipment used, condition of manholes, and any additional remarks.

Accurate, current maintenance history can protect the agency in case of a litigation challenge. During litigation, lawyers for both sides typically ask for maintenance history of specific items within certain dates. The CMMS must be capable of handling queries used to obtain the required information without too much manual input or data recreation.

Utilities currently using paper systems should consider computerized recordkeeping because of its lower long-term cost and ability to process information in seconds instead of hours or days. Using a CMMS gives a manager the ability to manipulate data in a tremendous variety of ways, from relational comparisons of different pipe materials and conditions, to automatic work-order generation based on television inspection done the day before. All are generated far faster than written records at far less cost over time. Some of these programs allow all operations (biweekly cleanings, staggered yearly maintenance, or repairs to the system or pumping stations) to be tracked and analyzed over time.

Once maintenance history and facility inventory data are available through GIS, the data also can be viewed in conjunction with other information such as, for example, a collection system's maintenance history overlaid with results from a hydraulic analysis. This compilation helps to prioritize capital improvement projects and to protect assets.

2.2.4.2 Complaints and Service Requests
Whenever a complaint is received from a system user, a service request needs to be recorded on either paper or in the CMMS. It is important that these requests be tracked and reviewed on a regular basis to identify trends and problem areas before they escalate. Guidelines for handling customer complaints must be developed and understood by all maintenance employees. Each agency should develop their own benchmarks to use in identifying areas for improvements. Various benchmarks have been developed for responding to complaints and other customer service issues, which are described later in this chapter. Charting customer complaints gives the manager more information for identification and prioritization of future capital projects. Service requests typically initiate a work order to address the problem.

2.2.4.3 Work Orders
The primary function of most CMMS applications is service order processing. All service orders must be geographically referenced and should provide information about the facilities involved. Location plays a key role in scheduling and tracking maintenance service orders and ultimately is included in a CIP. Work orders typically include the facilities involved; the problem reported; activities performed; materials, parts, and labor required; and job site information. Geographic information systems facilitate retrieval and use of this information by identifying specific facilities, locations, or attributes.

2.2.5   Integrating Geographic Information Systems and Computerized Maintenance Management Systems

Over the past few years, more utilities are converting to computerized systems to handle schedules, plans, specifications, engineering drawings, contracts, change orders, time sheets, and project communications and management. Computerized systems allow managers more time to manage projects; better and faster communications; less time and money spent producing reports, plans, and specifications; and a reduction in clerical staff.

Computers are playing a critical role in managing a utility's spatial and tabular as described above. The growing availability of geospatial data and GIS has given managers a valuable tool in planning, spotting trends, understanding the system, and presenting findings to staff, governing boards, and the public.

Integrating CMMS and GIS data with an existing system is one of the biggest challenges in creating a functional planning tool. This integration, however, can yield the greatest benefits. For municipalities, this includes looking at how data flows though the department, reviewing workflow processes, and potentially reengineering the information systems that support the organization. Because approximately 75 to 80% of the organization's information relates to location, proper implementation of a GIS can dramatically improve the effectiveness of this information. Automation and integration of work orders and linkage of the CMMS with the GIS allows maintenance supervisors, engineers and managers to statistically analyze complaints, recurring problems, and repair data. When linked to a GIS, data can be presented using custom queries to show pin maps or other thematic maps that show the type, frequency, and clustering of events. For example, a sewer's maintenance history can be overlaid with results from a hydraulic modeling analysis to rank potential capital improvement projects.

Combining the tabular (CMMS) data and spatial (GIS) data into an integrated tool allows managers the ability to more effectively plan and identify capital improvement projects that will

- Address recurring problems,
- Comprise projects that are spatially diverse across the municipality,
- Combine similar projects in close proximity, and
- Include information on surrounding construction projects.

2.2.6   *Inflow and Infiltration Identification*

The magnitude I/I in a collection system is a good indicator of overall system health and can be used to help identify and prioritize CIP projects. These extraneous flows can affect system performance, decrease capacity of existing infrastructure, and significantly increase the cost of wastewater treatment. Inflow and infiltration exist in all collection systems, and inspection crews are frequently tasked with identifying and tracking I/I parallel with condition inspection. Rehabilitation, refurbishment, and replacement of assets that are responsible for most I/I in a system is typically a large part of the overall CIP.

Inflow is extraneous stormwater or snowmelt runoff that enters a collection system through roof leaders; cleanouts; foundation drains; sump pumps; and cellar, yard, and area drains. Stormwater also may enter through cross-connections between a sanitary sewer and storm sewers and through defective manhole covers and frame seals.

Infiltration is water that enters a collection system from the ground through defective pipes, pipe joints, damaged service connections, or manhole walls. Infiltration most often is related to high groundwater levels but can also be influenced by storm events or leaking water mains.

Rainfall-derived infiltration and inflow (RDII) is I/I that occurs during and following a rain event. It has been recognized as a significant source of operating problems, often contributing to sanitary sewer overflows (SSOs) such as backups into homes and basements. The RDII also can cause serious operating problems at wastewater treatment facilities because it leads to increased flows, which can sometimes overload hydraulic or process capacity of plant facilities.

Inflow issues typically are easier to identify during wet weather because drains or sump pumps become active. These inflow sources typically are privately owned and need to be terminated by the property owner causing the discharge. In many instances, the property owners are not aware that they are discharging stormwater to the collection system.

Infiltration can be more difficult to pinpoint and more expensive to resolve than inflow. Although the amount of infiltration can be estimated with some accuracy by a good flow-monitoring plan, the actual infiltration sources can be especially difficult to find. They consist of underground defects that admit stormwater or snowmelt that has soaked into the ground. Though some leaks can be found during and immediately after wet-weather events, many will take days or weeks to find as groundwater levels increase enough to contact the collection system. Flow monitoring typically is used to find these areas of extraneous flows, which are then televised during periods of high groundwater levels to determine where the flows originate. After being cataloged, repairs can be made on a priority schedule based on estimated volume of infiltration.

The identification and quantification of I/I flows in the system typically is done by using testing techniques such as flow monitoring or portable instantaneous flow weirs. The use of CCTV, smoke testing, and dyes also can provide information on extraneous flows. Some large utilities have found it economical to keep chemical grouting and sealing equipment on hand during CCTV activities for spot repairs. It is important to catalogue all defects identified before making repairs; data can be used in future analysis and prioritization of similar pipe for CIP projects.

Inflow and infiltration data collected during inspection are used to identify and quantify specific sources of extraneous flows and determine which segments are affected. After significant wet-weather flow monitoring and instantaneous weir measurement data recording, flows are analyzed to determine where subsequent wet-weather and dry-weather I/I investigation work should be performed. The objective of an I/I investigation is to identify extraneous flow sources so that those pipes can be targeted in the CIP.

Field investigations typically include several tasks:

- Flow monitoring and instantaneous weir measurement to:
  - Determine wastewater and extraneous flowrates;
  - Determine collection system basins with excessive extraneous flows;
  - Identify collection system reaches with excessive extraneous flows; and
  - Quantify manhole extraneous flows.
- Internal manhole inspection to identify extraneous flow sources in manholes and tributary pipes.

- Wet-weather television inspections to determine extraneous flow sources and distribution between public and private property laterals.
- Dry-weather, CCTV inspections.
- Dry-weather smoke testing.
- Dry- or wet-weather dye water testing at suspected storm sewer direct and indirect connections to the collection system.
- Wet-weather surface inspections for manhole inflow sources.
- Building inspections for the identification of illegal connections such as roof leaders, sump pumps, and foundation drains during construction and remodels.
- Identification of cross-connections such as storm water drains.
- Small television camera inspections of service connections with extraneous flow or maintenance problems.
- Rainfall data collection.
- Work order review.

These investigations are necessary to identify and control extraneous flows that hydraulically overload some collection systems, pumping stations, and WWTPs. In most instances, severe extraneous flows will activate SSOs and cause backups that create health and environmental problems. Reduction of extraneous flows reduces the cost of wastewater conveyance and treatment, increases available system capacity, and reduces the need for parallel collection systems and flow equalization facilities or additional treatment capacity.

Hydraulic models described later in this chapter are a valuable tool in determining the effects of I/I on system components listed above and the benefits of its elimination.

2.2.7 *Flow Monitoring /Monitoring*

Knowing how gravity flow and pumping stations convey wastewater through the collection system is essential for understanding how the collections system functions on a day-to-day basis. Knowing the depth and velocity of wastewater moving through the system and determining what the theoretical volume should be based on the number and types of users can help determine what percentage is normal flow and what is attributable to I/I.

Obtaining accurate flow measurements is useful when developing a CIP. Flow measurements can help identify problem areas such as areas with high rates of I/I during wet weather. In addition to locating extraneous flows, measurements inform the system manager of the amount of wastewater being conveyed versus the capacity and estimated user volume of the system overall or in a given area. When development begins to tax capacities of pipes, flow measurement may be the only means available to regulate land use or system upgrade parameters. This information also is valuable for bypass pumping or for sizing a new pump for a lift station. Flow measurements can be used to determine I/I by comparing dry- and wet-weather measurements. In some cases, flow measurements are used to detect wastewater leaking out of the pipe and water entering it. As opposed to totalizer flow monitors—which typically are recorded daily and provide total flow quantities between monitor readings—continuous monitoring can be used for more detailed analysis. For example, continuous monitoring can be used to measure treatment flowrates from joining communities for billing purposes, and projected extraneous flowrates and volumes. Older monitoring facilities initially used depth-only monitors to measure flowrates, which can be inaccurate if submerged flow or backwater conditions occur or if debris buildup affects the primary device. Typically, these monitors were used in permanent or temporary manhole locations, providing a printed record on a circular chart on a daily or weekly basis. Advances in technology now provide monitors that can read depth and velocity simultaneously, providing greater accuracy during normal and surcharge flow conditions over longer periods of time. Data typically is stored electronically, or a telemetry system can be used to transfer information to a remote monitoring site.

Temporary monitoring is used to determine various flow conditions, including

- Dry-weather wastewater flowrates;
- Influence of extraneous flows;
- Flowrates for computer modeling;
- Peak and average daily flows for lift station design, bypass pumping, system upgrade, and interceptor sewer design;
- Available capacity for development;
- Location of bottlenecks;
- Project exfiltration rates; and
- Sewers with maintenance or capacity difficulties.

Many utilities use flow-monitoring devices to measure wastewater flows from sub-basins in their collection systems to assess the effects of I/I. Dry-weather, wet-weather, and rainfall-induced flow conditions are recorded. The dry-weather diurnal flowrates are used to identify base wastewater discharges and the presence of groundwater infiltration. Wet-weather flow conditions occur during rainfall and typically lead to the highest rates of flow in the collection system. Analysis of flow-monitoring data can be used to distinguish between wastewater flow components and identify where RDII flowrates and volumes are the highest, thus providing priorities for collection system rehabilitation.

Projected groundwater infiltration and RDII rates are determined through numerical and graphical comparisons of dry- and wet-weather flows. A graphical comparison of dry- and wet-weather flow patterns is an excellent method of projecting I/I rates and visualizing effects at the monitoring site. Figure 4.1 is a

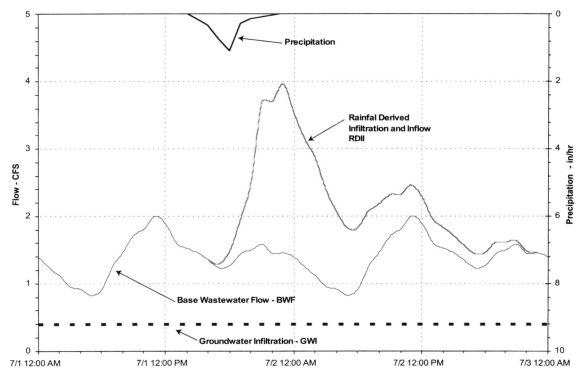

FIGURE 4.1 Typical combined sewer system schematic (courtesy of the City of Edmonton, Alberta, Canada, 2002) (cfs × 0.028 32 = m^3/s; in./hr × 25.4 = mm/h).

hydrograph (a plot of wastewater flow during a selected period of time) that projects RDII during a 24-hour period. A rainfall-induced flow plot is superimposed over a typical dry-weather flow pattern for the same day of the week (for example, Monday versus Monday) at the same monitoring site. The hydrograph notes the peak rate and volume of RDII during the analysis period. Through this type of flow comparison, the daily flow patterns or cycles at a particular monitoring site are matched. Any difference in the flow patterns is attributed to wastewater fluctuations or I/I.

Along with flow monitoring of extraneous flows, rain gauges should also be used to determine rainfall amounts, duration, and intensity. Comparisons between rainfall and flow monitoring determine the relationship of rainfalls to collection system extraneous flow responses.

2.2.8 *Hydraulic and Hydrologic Modeling*

Hydraulic and hydrologic modeling is a powerful tool that managers can use to evaluate the current and future state of the collection system. The material in this section provides a high-level description of the critical components of a hydraulic and hydrologic model. For more a more detailed discussion on this topic, refer to *Existing Sewer Evaluation and Rehabilitation* (WEF, 2008). This section uses excerpts from that manual for purposes of giving the wastewater collection system manager enough information to evaluate the required data and effort to use a hydraulic model for use in collection system management.

Hydraulic and hydrologic models traditionally have been used as planning tools and to efficiently evaluate alternatives and solutions. In addition, when merged with GIS and other applications, models can extend beyond planning projects and become an integral part of an infrastructure management toolbox. In this capacity, models can serve multiple objectives:

- Master planning;
- Capital improvement plan development;
- Regulatory compliance;
- Combined sewer overflow long-term control planning;
- Capacity, management, operations, and maintenance assessments;
- Inflow and infiltration evaluation;

- Design improvements;
- System O & M;
- Real-time control modeling; and
- Water quality/watershed studies.

The development of a model varies depending upon the intended objective and purpose of the model and the information required from it. Most municipalities will have two primary objectives:

(1) Develop a model to accurately simulate operation of the existing wastewater collection system and its ancillary structures and
(2) Develop a model that will become an operational and planning tool to guide ongoing maintenance activities and assist in the allocation of future wastewater collection system capacity to new development.

For capital planning, it is important to understand the level of sophistication of the model used and what assumptions are being made in the development phase. Some models can select more conservative projects to ensure that the system will be sized to handle the worst case scenario, but this may cost more than decision makers are willing to spend. Understanding the benefits and limitations of a model will allow managers to set an acceptable level of risk as they develop their long-term CIP.

2.2.8.1 Steady-State Simulation versus Dynamic Simulation
When looking for a hydraulic model, it is important to have a good understanding of the system and what is important to its operation. If the system routinely surcharges and pipes are serving as storage, then the model used must accurately be able to predict the shape of the hydrographs in both surcharged and non-surcharged flow conditions. Other factors that should be taken into account are whether it is necessary to simulate:

- Surcharge levels or flooding;
- Backwater effects in nonsurcharged pipes;
- Reverse flows in nonsurcharged pipes;
- Looped systems; and
- Modeling of ancillaries (e.g., pumping stations and regulators).

Where there are complex control systems or ancillary structures, special consideration should be given to the capability of the model to simulate these features. If the model is to be used as the basis for a collection system quality model, this will have to be considered in selecting software.

Existing steady-state models are not sufficient to develop this level of understanding. Although they have their place for some applications (design of new systems, for example), they are not suited for developing a detailed understanding of existing system operating characteristics required by most projects today. Dynamic modeling will provide

- Detailed understanding of collection system performance;
- Ability to project into future;
- Accurate understanding of I/I mechanisms;
- Basis for prioritization of improvements;
- Optimization tool, including real-time control alternatives;
- Integration of collection and WWTP models; and
- Regulatory defense.

Dynamic modeling of the collection system should be used to determine available system capacities and, from that, where the collection system can be used to provide interim capacity relief for the treatment plants. Further, dynamic modeling will allow the analysis of real-time controls for developing capacity augmentation measures that may include remote or other flow control measures and/or inter-basin transfers within a municipality's collection system.

2.2.8.2 Flow Development
For modeling purposes, it is useful to further define the components of flow in wastewater collection systems. Figure 4.1 displays a typical two-day wastewater flow hydrograph. Flow is plotted against the *y* axis on the left side of the plot and precipitation intensity is plotted against the inverted *y* axis on the right side of the plot. The wastewater flow hydrograph can be broken into three basic components:

- *Groundwater infiltration* enters a collection system typically through service connections and from the ground through defective pipes, pipe joints, connections, or manhole walls; may have seasonal variations but is typically treated as constant over the duration of a single precipitation event.

- *Base wastewater flow* enters the collection system through direct connections and represents the sum of domestic, commercial, and industrial flows.
- *Rainfall-derived infiltration and inflow* is the portion of the wastewater flow hydrograph above the normal dry weather flow pattern; it is the wastewater flow response to rainfall or snowmelt in a watershed.

2.2.8.3 Calibration/Verification

Model calibration is the process of adjusting estimated model parameters to match observed flows and water depths in the system. This step is essential for the creation of a model that accurately represents a collection system. In general, there are few parameters that can and should be adjusted; the final value should always fall within the range of reasonable values for that parameter. If a value outside of that range is needed to achieve a suitable match between model predictions and measurements, the model likely is not accurately representing the system. This is one of the reasons that inspections may be required: to check that actual system configuration matches available information. In general, calibration proceeds in the downstream direction. To accurately calibrate the hydraulic model, two efforts are needed: dry weather calibration and wet weather calibration. The goal of calibration is to simulate flow in the collection system within an acceptable range of observed flow data. Calibration targets also should take into consideration modeling objectives and availability/reliability of all data sources used to develop the model.

Typical tools used in the calibration and verification of hydraulic models include:

- Flow monitor data (dry and wet weather data);
- Rain data (point and radar);
- Facility information (i.e., lift station set points, run times, pump curves);
- Water usage;
- Land use; and
- Infrastructure asset information (location, diameter, length, etc.).

The required level of calibration and flow monitoring are determined after defining project goals and carefully reviewing existing information on dry- and wet-weather hydraulics. With adequate calibration, the models become consistent tools to assess the hydraulics of the existing collection system and to identify capacity concerns to be incorporated to a CIP.

2.3 Condition Assessment

Assessment of the collection system should be part of the daily activity of the maintenance staff. When determining the replacement priority of an asset, age is a valuable piece of information, but actual condition is much more useful. Everyone involved in field activity should gather information on system conditions as they perform their assigned tasks. For example, in addition to setting up and performing flushing and cleaning activities, operations staff could perform an above-the-ground manhole inspection at the receiving end while waiting for the equipment. This type of joint activity maximizes the use of operations staff time and also provides valuable information to engineering staff for CIP prioritization and I/I analysis.

Collection systems management is charged with maintaining the system and the investment it represents. An asset inventory tracked in a CMMS can provide the reporting and inventory analysis required by management. The data, if collected and stored properly, can also be used for planning and predicting system upgrades and future capital plans. A well-managed system should track several items:

- Population served by the system;
- Total system size in length of lines;
- Inventory of pipe lengths, sizes, and access points;
- System appurtenances, bypasses, siphons, overflows, diversions, and splits;
- System component installation dates and materials;
- Total theoretical capacity of the system;
- Total theoretical generation of wastewater by population;
- Volume customers;
- Industrial and commercial users with high discharge levels;
- Total current flow volume;
- Total length off-road easements and rights-of-way;
- Force main locations and lengths;
- Summary of pipe sizes and lengths of each;
- Number, location, and capacity of pumping stations;
- Number and location of siphons and inverted siphons;
- Number, location, and size of tide or flood gates;

Collection System Assessment and Capital Improvement Planning

- Equipment on hand (maintenance, excavating, and repair equipment);
- Number of field personnel available;
- Breakdown of pipe sizes and materials;
- Breakdown of interceptor, trunk, branch, and lateral lines maintained;
- Condition of physical line sections as available;
- Manhole inventory—location, size, type, and condition;
- Lift station components, including pumps and valves; and
- Complaint totals, including type, location, cause, and frequency.

There are many more detailed items that can be added, depending on the size, complexity, and topography of the system.

The Water Environment Research Federation (WERF) has developed protocols for assessing the condition of wastewater utility assets (AWWA Research Foundation and WERF, 2007). The research outlined a 10-step approach to developing a condition assessment program to align data collection and decision-support efforts:

Step 1. Document program drivers.

Step 2. Specify program objectives.

Step 3. Identify asset types to assess.

Step 4. Collate and analyze available data.

Step 5. Determine what assets to inspect, if any.

Step 6. Select inspection/assessment technique.

Step 7. Plan inspection program to minimize cost.

Step 8. Undertake asset inspection and other data collection.

Step 9. Analyze data and assess asset condition.

Step 10. Use condition assessment information for decision-making.

2.3.1 Surface Inspection

Some problems can be detected by visual inspection while walking or traveling along the rights-of-way of the collection system line, both in the street and in off-road easements. Sidewalk irregularities; cracked, settled, or dipped pavement; ponding over manholes; or depressions along the path of the pipe can be indicators of leaking or collapsing pipes. If a joint is bad or a pipe is broken, wastewater may wash away the surrounding soil and create a cavity beneath the surface. Sometimes, the weight of the overlying soil is enough to cause collapse and a depression at the surface. In easements, these depressions can be seen as flooded or sunken areas along the pipe route, especially after a significant wet-weather event.

Cavities under the road surface are often reported as a depression in the street. Such a report may go to a street repair crew, and the depression would be filled without further investigation. To avoid this, the cooperation of the street repair department is essential. Cavities that are left unattended will enlarge, and the weight of a vehicle can be enough to cause a road collapse. Depressions (settling) can also be the result of inadequate compaction during construction or subsequent utility work.

Manhole inspections are another source of surface surveillance information; problems can be reported on a standard form. The form notes items such as evidence of high water or surcharging; missing bricks, concrete, or rungs; and water entry through the barrel, riser, frame, or cover of the manhole. Broken, raised, or sunken manhole frames or covers can be noted and repaired quickly.

Manholes periodically are examined for any sign of bedding material in the invert that suggests a breech or collapse in the upstream line. Deteriorated steps may indicate the presence of corrosive gases, possibly hydrogen sulfide. Information on the completed forms is transferred to the computer and loaded to the data file to be used later for listing priority, difficulty, or common traits.

In both surface and underground inspection, evidence of water entry should be noted for future consideration or solution. Adding this information to the system inventory will create a far more comprehensive database from which hydraulic modeling can be used to predict system upgrades or repair needs.

2.3.2 *Closed-Circuit Television*

Some older methods of visually inspecting pipe are still used, such as manhole entry to lamp portions of collection system pipes, entry into large pipe for visual inspections, and the use of mirrors on adjustable poles. However, most internal inspection is performed by the use of CCTV systems. Televising of pipes eliminates the need of personnel entry into manholes and can provide a video record of conditions in existing pipes.

Television inspection systems operators should be trained in equipment operation, collection system defect classifications, and data recording procedures. Any defects in the pipe should be noted on the video, and the appropriate defect code should be noted in the software that typically accompanies inspection equipment. Most CCTV inspection software now has the ability to link video with inspection results, which give the user the ability to jump directly to the defect. These defect records should be documented according to the coding or rating system adopted by the investigating agency. There are numerous coding systems in use, ranging from basic, handwritten codes with 10 to 20 classifications, to computer applications that contain numerous code classifications and matrixes. The National Association of Sewer Service Companies has developed a national standard for assessments called the Pipeline Assessment Certification Program (PACP). An increasing number of collection system operators are moving to the PACP coding system, which is also supported by many of the inspections software programs.

Computer systems allow storage of digital video and inspection results for rapid processing and cross-referencing that can be used to evaluate and rank large quantities of data. The agency should select a coding system that is appropriate for its particular applications and train operators to properly use it to avoid confusion or errors. Some utilities use the operator's rating system to help determine priority levels. If a rating scale is used, operators should be trained in making assessments because most ratings will be subjective to some degree.

Data management is required after surveillance records are collected. Some utilities continue to use a paper system that splits information into areas of responsibility so that it can be reported to various departments. Others are able to tie the inspection data directly into the CMMS for use in tracking costs, time, and condition of pipeline. In addition, by tying directly into a CMMS, the information collected by field operations staff can be made available throughout the utility. This sharing enables the work-order management system to issue repair requests based on severity and priority of the problem identified in the field. This procedure takes less time and allows the data to be handled only once. Priorities and repair or upgrading projects are clear so that long-term budgets can be used and crisis situations avoided.

2.3.3 Smoke Testing

When evaluating I/I in the system, the inspection crew frequently uses smoke testing as a means to identify direct inflow connections and system failures that should be repaired or addressed in a CIP. Smoke testing involves a specially mounted blower that forces air into a manhole and pipe and smoke "bombs" of assorted duration. The smoke generated is nontoxic, has no odor, and typically is foggy-white in color. The smoke is forced by the blower into the line and seeks the path of least resistance to exit the pipe. Under average circumstances, the structures in the section of pipe being tested will release this smoke from the vent stacks. If gutter drains are connected to the service line, smoke will appear from the downspouts. As a precaution, when engaging in smoke testing, descriptive information should be supplied to every building on the line before the testing. Most utilities use a simple door hanger that informs the public of the work being done to allay any concerns of toxicity of the smoke.

When conducting smoke testing operations, the fire department should be notified if there is a sump pump or floor drain connected to the service line through which the smoke will exit the well. People could mistake the smoke for a fire. It is important to note that smoke testing typically is not effective in locating all sump pump connections. Some pump systems incorporate "P" traps that block the smoke or are in basements that are not viewable from the street or lawn areas.

Most property owners are not aware of illegal connections to the system. In such cases, when smoke is viewed exiting from anything but the vent stack, a photo should be taken and shown to the owner and attached to the inspection report for future reference. Plans can then be made to terminate the connection, typically at the owner's expense. In some areas, local, state, or county funding may be available to help defray the cost to system users of disconnecting these sources.

Smoke testing also can show cross-connections to storm sewers and other underground leak points. Smoke will exit from storm drains or appear out of the ground, indicating a problem. These problems will require visual inspection of the pipe to determine where the opening or cross-connection is and can sometimes be difficult to trace because the smoke can drift between underground pockets and appear far from the actual release point.

Building inspections are another means of inflow control. Though time-consuming, inspections may be the only way to find illegally connected sump pumps and floor drains. Building inspection can be conducted while smoke testing is ongoing, with few property owners not allowing access to structures. Some utilities work with the local fire and building departments when they inspect private and commercial property, asking them to note any pumps or obvious drainage that goes to the service connection.

2.3.4 *Trunkline Inspection*

Trunklines and large-diameter pipe typically are inspected and maintained on a separate schedule and with different equipment than smaller collection system lines. Trunklines are a critical component to a collection system in their carrying capacity and in the potential for catastrophic failure. Special attention to these pipes is important for maintaining a safe and operational collection system.

Methods of inspection for large-diameter pipe include CCTV and the use of sonar to image the interior of the pipe. The advantage of sonar is the ability to identify the wall thickness and amount of debris deposited on the pipe floor. In a large-diameter pipe inspection, it is important to identify the critical crossings and to prioritize based on the available funds. These sensitive crossings could include highway, railway, and water bodies. Results from these inspections should be recorded and prioritized in a CMMS for use in determining capital projects. Typically, replacement of a trunkline has a large financial effect on a municipal budget. By regularly inspecting these trunks, potential problems can be identified early and added to the CIP to be budgeted in later years.

2.3.5 Lift Stations

An equipment-oriented maintenance management system can reduce the time and cost of pumping-station management. Computer-based maintenance management systems offer additional advantages. The steps required to order a replacement pump part illustrate the large amount of information that must be stored, tracked, and integrated:

- Issuing a work order,
- Obtaining equipment nameplate and master parts list data,
- Checking spare parts inventory, and
- Purchasing or replacing the part in inventory.

In addition to maintaining the facility, using a CMMS will assist the manager in planning for and funding capital improvements. A CMMS will allow the manager to track wear and tear, identify problem parts or equipment, and prioritize large capital expenses. This can be a time-consuming task when performed manually. For computer-based systems, information can be stored in databases and retrieved quickly to expedite the process. There are several functions of these systems:

- Storing all relevant equipment nameplate data,
- Generating work orders,
- Tracking inventory and warehousing operations,
- Tracking maintenance history,

- Generating purchase orders,
- Keeping records of maintenance activity, and
- Reporting.

Computer-based budgeting and accounting functions are necessary to optimize budget control and analysis and to evaluate maintenance program performance. Analysis of flow data, rainfall, pump running times, and power consumption are all done on a routine basis to assist in data management.

3.0 CAPITAL PLANNING

A CIP is a multiyear planning instrument used to identify capital projects that are needed to increase the capacity of the system and to rehabilitate or replace infrastructure that is deteriorated or damaged from overuse or inadequate maintenance or repair. In addition, a CIP aims to coordinate the financing and timing of these improvements in a way that maximizes the return to the public.

Increasingly, professionals and policymakers are becoming aware that infrastructure exists as an interconnected system and that efficient functioning of the system is crucial to the environment, economy, and quality of life. The wastewater collection system, which includes sewers, pipelines, conduits, pumping stations, force mains, and other collection system-related facilities, is one of the largest infrastructural assets a community owns. Both public and private wastewater companies are being challenged to meet new state and federal regulations, increased operating costs, and the expectations of customers for higher standards, while facing a deteriorating infrastructure and funding constraints. Collection systems degrade continuously, and a plan is necessary to effectively manage this degradation and prevent collection systems failures caused by pipe aging. Deferred maintenance will only increase costs for later generations and may cause catastrophic results.

The legal context under which municipal utility decisions have historically been assessed is that of utility law. The standard holds municipalities to a duty similar to the one imposed on a privately owned public utility; that is, that service be supplied to all customers in similar positions to those already being served. Courts analyze municipal decisions relating to extensions of service on the basis of reasonableness and whether or not the extension discriminates against any individual or group of property owners.

Although utility law currently is the common law rule applied in most courts, new legal emphasis is being placed on municipal planning for growth and regional coordination of planning. Several states have complex legislation that links capital improvement planning directly to comprehensive plans. Some courts have recognized the value that comprehensive planning considerations have to utility decisions. Whether or not a community is under the stress of rapid growth, the incentive to plan for future growth is present in nearly all of the states.

A CIP provides an excellent basis for making the link between growth management plans and utility policy. It allows a community to plan for future development by realistically assessing utility obligations, while formulating its policy to reflect the community's growth goals.

An integrated CIP should carefully plan for extension and expansion of collection system systems. An increasing number of municipalities are requiring the costs of system growth to be placed on new users. As financing of new service and system rehabilitation becomes a budgetary burden on localities, the CIP can provide a sound basis for equitably and legally distributing the costs of system expansion.

In light of continuing changes in the legal importance of planning for growth, it is clear that state courts will have to address the conflict between utility law standards and the increasing need for municipal planning and growth management. When dealing with a challenged municipal utility policy, courts will have to determine a basis for that policy and may run up against conflicting objectives for collection system and water provisions. Objectives must balance growth management principles with what can be legally recognized as "utility-related" rationale. If such a policy is clearly outlined by a CIP that integrates utility cost and revenue projections with local growth management objectives and land use design, then a reasonable basis for judicial sanction will become clear.

Utility providers will not be able to operate in a vacuum. Good planning practices bring together affected stakeholders to determine how they want to prioritize their limited resources among level of service, the environment, and public health concerns. These concerns vary from community to community and from system to system. Planners and designers must extend their thinking beyond design tables and standardized approaches. *Guide to Managing Peak Wet Weather Flows in Municipal Wastewater Collection and Treatment Systems* (WEF, 2006) helps stakeholders understand how others have approached the planning process and evaluated the challenges that utility providers face. As service provision becomes more expensive and more important to metropolitan growth, an integration of values must form the basis for future utility extensions, rehabilitation, and financial policy. A CIP is based on the standard of service required by the customer. The CIP provides a means to protect, maintain, or improve the asset value of a collection system with planned repair, rehabilitation, or replacement on predicted deterioration of the system. In either a private or public utility, key information is needed to manage costs through asset-management planning (Kingdon, 1995). This information includes current conditions and performance of assets, operating costs, and financial position including revenues, balance sheet, and cash flow; required and anticipated future levels of service; and methods of measuring and monitoring performance of the system.

The primary goals of collection systems operation are to prevent public health hazards and to transport wastewater uninterrupted from source to treatment facility. The goal of a CIP is to predict the costs necessary to protect, maintain, or improve the value of this asset (the collection system) while providing the level of service desired. To meet these goals, the following factors must be addressed:

- Determining existing conditions,
- Setting future goals,
- Attaining future goals, and
- Tracking progress.

3.1 Master Planning

A master plan is a key component for developing a comprehensive CIP. Master plans typically are developed and updated every five years to address system deficiencies and future growth. The basic goals of a master plan include:

- Developing a long-term vision and strategy for management of the wastewater system.
- Identifying existing and future system deficiencies.
- Identifying and prioritizing system improvement projects to address current and future needs.

Master plans typically view the system through a predefined "buildout" year, often determined as part of a general plan or other community or utility master plan document. Most master plans first identify existing deficiencies in the system based on current system conditions and flows. Examples of existing conditions include capacity issues, severe I/I, overflows, and areas of known maintenance problems. Based on these existing systems analyses, projects are identified and added to the CIP. The second step is to reevaluate the system during build out and identify what components of the system would need to be replaced or upsized for future flows. A hydraulic model as described in previous sections is a useful tool in this type of analysis. This analysis identifies the incremental improvements needed to address the future flows and growth to the system. This long-term approach provides a detailed roadmap for significant improvements needed for the system through build out and to estimate the funds needed to implement them. Typically projects identified in a master plan become the "base list" of projects when developing a CIP. As described later in this chapter, it will be important to identify the priority and timing of projects as they are incorporated into the CIP.

3.2 Capital Improvement Plan Development

A typical city charter requires that the city manager prepare and submit a CIP for the five-year period following the new budget year as part of the annual package. The CIP is a planning document and typically does not authorize or fund any projects.

Prior to the city manager bringing forward a CIP, preliminary reviews typically takes place to verify the accuracy, sustainability, and priority of projects selected. In most cases, this is done as a progression from the section (i.e., wastewater) to the department (i.e., utilities) to the governing board.

3.2.1 *Ratings*

To optimize decision-making and to help prioritize all potential solutions to a problem, it is important to incorporate the information that was collected using the methods described above and grouping them into three categories:

- Condition,
- Performance, and
- Risk.

This information can be converted into ratings to compare one scenario against another. Although cost is an important factor in identifying a solution, it is not the only factor that should be considered when developing a CIP. There are many ways to develop a scoring scenario, including the one described here. Each municipality will be different and should modify its approach to fit the agency's established business model.

The scoring procedure shown in Figures 4.2 and 4.3 is a method that can be used to look at various aspects of an asset and normalize them to make prioritization decisions. The example takes into consideration condition, performance, and risk, which

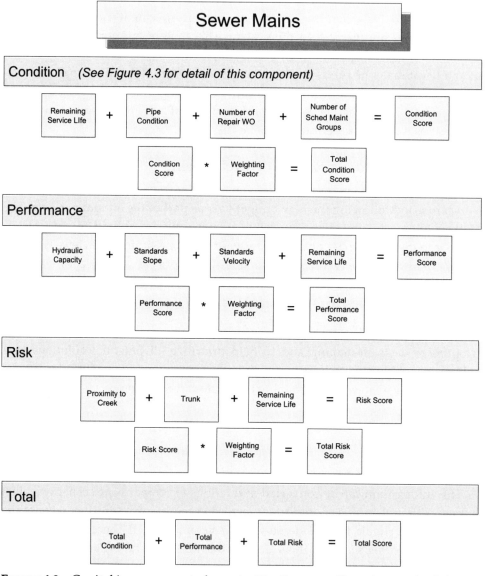

FIGURE 4.2 Capital improvement plan prioritization overall scoring methodology (WO = work order).

Collection System Assessment and Capital Improvement Planning

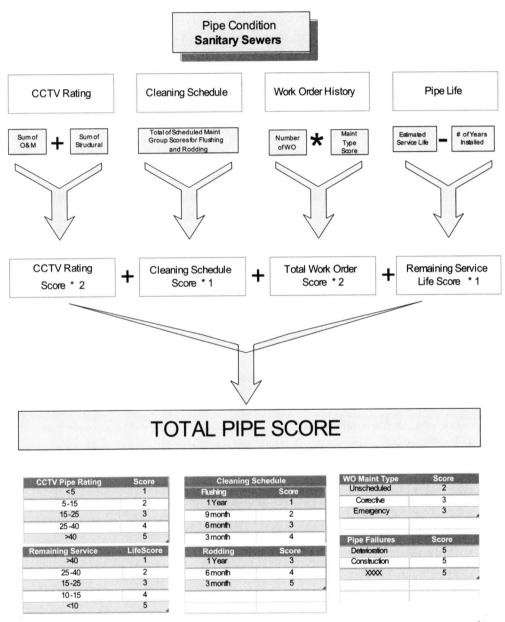

FIGURE 4.3 Capital improvement plan prioritization for sanitary sewer pipe condition (WO = work order).

are described in more detail below. By applying this type of methodology, every asset can be given a score (weighted appropriately based on each utilities' defined level of service). Once each asset has a score, GIS can be used to group the assets into projects resulting in a total score; Figure 4.4 includes this total project score. Having this score provides backup and justification for why one project is more important than another when presenting to management or the agency's governing body.

3.2.1.1 Condition

Condition scores are derived from the various tools and condition assessment methods described in previous sections. For example, if a pipe had sufficient capacity but most joints were offset or had excessive root intrusion, the condition score would reflect poorly. Examples of condition scoring components are:

- Remaining service life;
- Condition of pipe (based on physical inspections);
- Number of repair services requests for particular assets (tracked in a CMMS); and
- Required cleaning or maintenance frequency.

An example of how these components could be combined to give an overall pipe rating for use in prioritization of CIP projects is shown in Figure 4.2.

3.2.1.2 Performance

Performance scores are derived from the various tools and performance assessment methods described in previous sections. An example of a poor performing pipe would be a pipe that was unable to satisfy the level of service expected or set by the agency. A pipe could be in pristine condition, but if it cannot handle the design flow, its performance score would be poor. Examples of condition scoring components are:

- Hydraulic capacity;
- Design standards (slope, velocity);
- Flow characteristics; and
- Performance with build out flows.

3.2.1.3 Risk

Identifying what risk is involved, will help identify what exposure the utility will have and help determine funding priorities. If a particular asset has a high risk to the

Title	Status	Available	Y1	Y2	Y3	Y4	Y5	Future	Scope	R M P U E S L C I P A L U F B U I L L D I N G S E L E C I N G C S L P C O R R N E T A O R F Y S W I T A T E R T D V W P C S	Calc Priority	AM Priority
Colorado Blvd - Spring Creek to Sonoma									Sum Priority: 29			
TMP47 WM Replace: Colorado Blvd	PL	$572,464	$0	$0	$0	$0	$0	$0	Replace 1500' of 6 inch cast iron water main with 8 inch PVC. The pipe is in poor condition with 4 main breaks in eight months. Replacing this main will reduce maintenance costs and increasing the size of the main will improve fire flow.	☐☑☐☐☐☐☐☐☐☐☐☐	26	0 Field Crew Priority #2
8304 SM Replace: Colorado Blvd - Spring Creek to Sonoma	NEW		$0	$0	$500,000	$0	$0	$0	This project will replace approximately 1500' of existing VCP sewer main in Colorado Blvd. between Sonoma Ave and Spring Creek Drive with new 8" PVC. Replacing VCP sewer mains reduces inflow and infiltration and reduces maintenance costs.	☐☐☐☐☐☐☐☑☑☐☐☐	3	2
Midway/Magowan Area Phase I									Sum Priority: 24			
7608 WM Replace: Midway/Magowan Area Phase I	CO	$572,464	$0	$0	$0	$0	$0	$0	This project will upsize 4" cast iron mains in Midway Drive and Magowan Drive from Farmers Lane to Short Road in conjunction with a sewer project in the same area. Replacing the old mains with 8 PVC mains will increase available fire protection flow and reduce maintenance costs.	☐☑☐☐☐☐☐☑☐☐☐☐	2	0
7962 SM Replace: Midway/Magowan Area (VCP)	CO	$962,205	$0	$0	$0	$0	$0	$0	This project is the first phase of a four phase project in the Montgomery, Sonoma, Farmers Ln area. This portion will abandon 1875 feet of backyard VCP sewer main and will be replaced with new 8 inch PVC sewer main to be constructed in Midway Drive and Magowan Drive from Farmers Lane to Short Road. The backyard mains have dips, sags, intense root intrusion, and are difficult to access. Approximately 91 lots will need on-site plumbing to move their sewer laterals. Lots facing Montgomery Dr and Sonoma Ave will be replumbed to sewer to those mains.	☐☐☐☐☐☐☐☑☑☑☐☐	27	3 Field Crew Top Ten priority #3
Pierce/2nd/Montgomery									Sum Priority: 29			
8341 WM Replace: Pierce St/2nd St/Montgomery Dr	NEW		$0	$0	$300,000	$0	$0	$0	Replace approximately 1160 feet of existing 6 inch cast iron and asbestos water pipe with 12 inch water main in conjunction with Utilities sewer replacement project. Current City policy is to replace a main 10 inch mains in commercial areas. Replacing the main with the sewer project will negate the potential for street cuts and disruption to the neighborhood.	☐☐☐☐☐☐☐☑☐☐☐☐	10	0
8808 SM Replace: Pierce St/2nd St/Montgomery Dr	NEW		$0	$0	$550,000	$0	$0	$0	Replace approximately 830 feet of existing 8" VCP and upsize to 8" in Pierce St and 2nd St. Replace approximately 500 feet of 12 inch asbestos sewer pipe and approximately 345 feet of 12 inch VCP pipe and upsize to 18 inch sewer. This project will eliminate a second 6" VCP sewer main in 2nd St below the existing 12 inch sewer pipe. The project will reduce infiltration and lower maintenance costs. This project includes all of BO-07 2nd St Sewer Main, from Montgomery Dr to Brookwood St - Replace existing pipeline with 8&4' of 15" pipeline. This project includes all of EC-01 Pierce Street Sewer Main from 2nd to 3rd - Replace existing pipeline with 233' of 8" pipeline at a new slope.	☑☑☐☐☐☐☐☑☑☐☐☐	19	4 Field Crew Priority # 8
Brookwood @ Sonoma and 3rd St Water Main Replacement									Sum Priority: 27			
7633 WM Replace: Third St - Brookwood to Pierce	CO	$125,380	$850,000	$0	$0	$0	$0	$100,000	This project will replace the 4" water main in Third Street between Brookwood and Pierce Streets with a new 12" PVC main in conjunction with a sewer main renewal in the same area. Upsizing the water main to 12 inch mains in commercial areas will improve fire flow and bring the main to current City Standards for this commercial/multifamily residential area.	☐☐☐☐☐☐☐☑☐☐☐☐	2	0
8797 SM Replace: Brookwood at Sonoma	CO	$365,782	$0	$0	$0	$0	$0	$0	This project will replace approximately 500' of existing VCP sewer main in Brookwood Avenue just south of Sonoma Ave. Television records of this main show cracks and offsets throughout the pipe. Replacing the main will reduce inflow and infiltration and reduce maintenance costs. PW Overlay scheduled 2006.	☐☐☐☐☐☐☐☑☑☑☐☐	6	3
8757 SM Replace: Third St - Brookwood to Pierce	CO	$139,177	$0	$0	$0	$0	$0	$0	TV logs show roots and cracks throughout this sewer line. Approximately 925 of 6" VCP sewer main will be replaced with new PVC pipe. Replacing this main will reduce inflow and infiltration, and will reduce maintenance costs. The 4" water main in the same street will be upsized in conjunction with this project.	☐☐☐☐☐☐☐☑☑☐☐☐	19	2 Field Crew Priority # 8
Spring St Water & Sanitary Sewer									Sum Priority: 25			
7605 WM Replace: Spring St - 4th to Pacific	CONT	$96,766	$0	$0	$0	$0	$0	$0	This project will replace approximately 1950 of 4" water mains with new 8" water mains in conjunction with a sewer main renewal project in the same street. Replacing the water main will improve fire protection flow, and reduce maintenance costs.	☐☐☐☐☐☐☐☑☐☐☐☐	12	0
7688 SM Replace: Spring St - 4th to Pacific (VCP)	CONT	$22,578	$0	$0	$0	$0	$0	$0	This project will replace approximately 2500' of 6" VCP sewer pipe with new 8" sewer. It replaces old, deteriorated, flat pipe, and redirects flows to the new Benton Street Trunk sewer per the McDonald/College Neighborhood Sewer Study. Replacing these mains will reduce inflow and infiltration and reduce maintenance costs.	☐☐☐☐☐☐☐☑☑☐☐☐	13	1

FIGURE 4.4 Typical capital improvement plan layout and components (ft × 0.3048 = m; in. × 25.4 = mm; PVC = polyvinyl chloride; VCP = vitrified clay pipe).

agency or community if it were to fail, additional weight should be given. A pipe in poor condition that is low risk should be evaluated differently than a pipe in poor condition that is high risk for the agency. Examples of high business risk exposure would include:

- Assets at the end of expected life cycle;
- Proximity to sensitive areas (highways, creeks);
- Absence of redundancy or backup for a specific asset that could result in high cost of failure;
- Poor reliability scores; and
- Imminent significant failure mode of capacity or level of service.

Risk also is defined as the relationship between the consequence of failure and the probability of failure. More details on how this process is used in project validity can be found in WERF's strategic asset-management tool Sustainable Infrastructure Management Program Learning Environment.

By bringing together these various scoring elements, it is possible to develop a total score using weighting factors to satisfy particular utility goals and concerns. An example of how this process works is shown in Figure 4.2.

3.2.2 Validation

Once the projects and assets have been identified as candidates for rehabilitation, renewal, or replacement, it is important to validate the project. Questions to ask about a project include:

- Should the project be eliminated?
- Can the project be deferred?
- Can maintenance activities be adjusted or modified?
- Can the operation of the asset be changed?
- Is there a non-asset solution?

Answering these questions will help the utility determine if the specific project is the right solution, at the right cost, scheduled at the right time, and for the right reasons. A lot of work has been done to develop methods for assessing the validity of a project. One way is to develop a business-case evaluation that looks at all aspects of the project to ensure the right solution is being proposed. In the past, utilities looked only at the financial effects of a project. This is changing as more utilities are using the triple bottom-line method, which takes into account the financial, social, and environmental effects of the project.

3.2.3 Replacement/Rehabilitation Goals

Replacement goals are one way a municipality can identify the investment required to reach goals identified in the CIP or by governing boards. A replacement program helps facilitate the long-term budgetary planning and helps the municipality plan a gradual replacement of the aging collection system infrastructure in a way that is palatable to management, governing boards and the public.

A basic method for determining a replacement goal is to identify the pipe age and serviceability as described previously. Where pipe material or data is unknown, assumptions will need to be made based on location and interviews with field operations staff. In this basic method, replacement because of age and serviceability is assumed to be required at the end of a pipeline's average service life based upon industry standard and past experience by operations and engineering staff for a particular municipality. Using this method, it is possible to estimate the quantity of pipe that will exceed service life by a chosen date. This can then be converted into a replacement rate. Typically, municipalities will need to have a more aggressive replacement rate initially to bring the system up to a point where a more stable and palatable rate (i.e., 1%) can be applied in future years to accommodate pipe with a longer life cycle (i.e., more than 100 years).

3.2.4 Industry Benchmarks

Benchmarking is a process that management can use to gauge performance of a collection system and to define the goals and priorities of a CIP. Many large corporations have used benchmarking; the fundamental principles affecting productivity, profitability, or organization are as valid for wastewater collection systems. The benchmarking process is straightforward, as shown in the following six steps:

(1) Identify benchmarking functions,
(2) Identify performance criteria,
(3) Identify best-in-class utilities and measure best-in-class performance,
(4) Measure system performance,
(5) Close the gap between measured system and best-in-class performance, and
(6) Implement changes.

When identifying benchmarks, those functions that can be measured easily and that allow managers to exert influence through changes in policies and procedures should be targeted. Identification of performance criteria should include consideration of both financial measures (such as return on assets) and strategic measures (such as customer-perceived quality). When performance is measured, the organizations being compared should have the same variables. These performance indicators are sometimes needed for wastewater collection systems to justify budgets. Any improvement plans to close the gap between performance of comparable utilities must be weighed against the cost and investment required to achieve results. Finally, implementation of any changes should consider the time needed to achieve results and include a system to monitor and reassess results.

Benchmarking is defined as a method of comparing a utility or operation with a selected group of similar operations. The American Water Works Association and the Water Environment Federation have developed a quality service program called QualServe (http://www.awwa.org/Resources/utilitymanage.cfm?ItemNumber=3765&navItemNumber=1586). The indicators developed were designed to help utilities improve their operational efficiency and effectiveness by assessing their own utility against a standardized benchmark. Another goal of benchmarking is to provide an expanding inventory and condition assessment over time. The results can then be used for prioritization and justification of a CIP.

3.2.5 *Prioritization*

This chapter has described various methods and tools used to determine the condition of a collection system. It is important that a manager take all this information and prioritize it in a way that will produce a capital plan to:

- Improve water quality by reducing overflows from combined sanitary/stormwater sewers,
- Reduce costs by minimizing required emergency responses and repairs,
- Reduce costs by improving management and extending system life,
- Develop sufficient system capacity for existing and future growth,
- Substantially reduce the amount of I/I entering the system, and
- Identify the most cost-effective method for addressing system needs.

To help the manager prioritize the vast amount of information identified using the tools and methods described in this chapter, a decision matrix should be developed.

Collection System Assessment and Capital Improvement Planning

A decision matrix identifies system metrics that are important to the utility and assigns a weighting factor according to their respective value identified by management and staff. Metrics that should be included in developing a decision matrix include:

- Capacity deficiency,
- Inflow and infiltration,
- Known O & M problems,
- Design flow,
- Project cost,
- Project coordination,
- Service life,
- Required by regulation, and
- Identified in master plan.

3.2.6 Final Product

After performing a system analysis using the tools described above and identifying a plan incorporating master planning, prioritization, and agency goals, it is time to develop a plan to communicate the CIP to staff, governing body, and the public. A well-organized, complete CIP will include the following elements:

- Project name and identification,
- Budget estimate,
- Funding year,
- Priority rank (based on risk, condition, and performance), and
- Description of project.

Although every CIP will have unique layouts and varying information, the elements above represent a standard of best practices for a typical CIP. An example of a CIP with these elements is shown in Figure 4.4. The CIP is a planning document and does not authorize or fund any projects.

A significant component of this plan is the budget estimate and funding year. This information is used to help determine the most appropriate funding strategy to accomplish the plan. More description on this process can be found in Chapter 8, Budgeting and Financial Planning.

4.0 REFERENCES

AWWA Research Foundation; Water Environment Research Foundation (2007) *Condition Assessment Strategies and Protocols for Water and Wastewater Utility Assets;* Water Environment Research Foundation: Alexandria, Virginia; AWWA Research Foundation: Denver, Colorado.

Kingdon, W. D. (1995) Water/Engineering and Management. Water Research Center.

Water Environment Federation; American Society of Civil Engineers (2008) *Existing Sewer Evaluation and Rehabilitation,* 3rd ed; Manual of Practice No. FD-6; ASCE Manuals and Reports on Engineering Practice No. 62; McGraw-Hill: New York.

Water Environment Federation (2006) *Guide to Managing Peak Wet Weather Flows in Municipal Wastewater Collection and Treatment Systems.* Water Environment Federation: Alexandria: Virginia.

5.0 SUGGESTED READINGS

American Water Works Association (AWWA) *QualServe.* http://www.awwa.org/Resources/utilitymanage.cfm?ItemNumber=3765&navItemNumber=1586 (accessed October 2008).

Harlow, K. (2007) Beyond Risk: A Community Return on Investment Approach to Pipe Rehabilitation, *Underground Infrastructure Management.* November/December. pp 38–40.

Kirby, R.; Cummings, R.; DiTullio, B. (2007) Asset Management in Practice: Lessons Learned from Hillsborough County. *Underground Infrastructure Management*, September/October, pp 25–29.

Lafferty, A. K.; Lauer, W. C. (2005) *Benchmarking Performance Indicators for Water and Wastewater Utilities: Survey Data and Analyses Report;* American Water Works Association: Denver, Colorado.

Linschoten, G. (2007) Lifecycle Analysis and Capital Planning Using Asset Condition and Deterioration. *Proceedings of the 80th Annual Water Environment Federation Technical Exposition and Conference* [CD-ROM]; San Diego, California, Oct 13–17; Water Environment Federation: Alexandria, Virginia.

National Association of Sewer Service Companies (NASSCO) Home Page. www.nassco.org (accessed February 2008).

Water Environment Research Foundation (WERF) *Sustainable Infrastructure Management Program Learning Environment.* http://www.werf.org (accessed October 2008).

Chapter 5

System Design Considerations

1.0	INTRODUCTION	99	2.0	DESIGN GUIDELINES	103
	1.1 Regulatory and Environmental Requirements	100		2.1 Design Process Checklist	103
				2.2 Project Management	104
	1.1.1 National Pollutant Discharge Elimination System Permits	100		2.3 Using Consultants	105
				2.4 Routing Considerations	105
	1.1.1.1 Sanitary Sewer Overflows and Combined Sewer Overflows	100		2.4.1 Assessing Aboveground and Belowground Interference	105
				2.4.1.1 Surveys	106
				2.4.1.2 Potholing	106
	1.1.1.2 Capacity Management, Operations, and Maintenance	100		2.4.2 Geotechnical	106
				2.4.3 Traffic Control	106
				2.4.4 Permanent Access	107
	1.1.2 System Conveyance Design Rates	101		2.4.5 Public Effects	107
				2.4.6 Rights-of-Way and Easements	108
	1.1.2.1 Risk-Based Design Options	102		2.5 Hydraulic Analysis Using Modeling	108
	1.1.2.2 Wet-Weather Facilities	102		2.6 Pumping Stations	109
	1.1.3 Design Approvals	102		2.6.1 Package Stations	109
	1.1.4 Environmental Issues and Compliance	103		2.6.2 Other Types of Stations	109
	1.2 Permitting	103			

(continued)

2.7	Pipelines and Joint Materials	110	2.8.3 Structural Repairs	124
	2.7.1 Gravity Pipelines	110	2.8.4 Technologies for Design	124
	2.7.1.1 Sizing Considerations	110	2.8.5 Inflow and Infiltration	125
	2.7.1.2 Materials	111	2.9 Manholes	125
	2.7.1.3 Backfill and Bedding	112	2.9.1 Spacing	126
	2.7.1.4 Grades	114	2.9.2 Materials	127
	2.7.1.5 Inflow and Infiltration Allowance	114	2.9.2.1 New Construction	127
	2.7.2 Pressure Pipelines	115	2.9.2.2 Rehabilitation	127
	2.7.2.1 Types of Pressure Systems	115	2.9.3 Channels	128
	2.7.2.2 Sizing Considerations	116	2.9.4 Covers	128
	2.7.2.3 Materials	116	2.9.5 Steps	129
	2.7.2.4 Backfill	117	2.9.6 Drop Manholes	129
	2.7.2.5 Restrained Joints	117	2.9.7 Connection between Manhole and Sanitary Sewer	130
	2.7.2.6 Blowoff and Air Valves	118	2.10 Manhole and Pipeline Testing	130
	2.7.3 Odor and Corrosion Control	119	2.10.1 New Construction	130
	2.7.3.1 Sulfide Control Design Considerations	120	2.10.2 Rehabilitation	131
	2.7.3.2 Corrosion Protection Design Considerations	122	2.11 Inverted Siphons	132
2.8	Rehabilitation	123	2.12 Sanitary Sewers above Grade	133
	2.8.1 Condition Assessment for Design	123	2.13 Crossings and Tunnels	133
	2.8.2 Sewer System Assessment Protocols	123	2.13.1 Stream Crossings	133
			2.13.2 Other Crossings	133
			2.14 Service Connections and Disconnections	134
			2.15 Value Engineering	135
			2.16 Specification Writing	135
			3.0 REFERENCES	136

1.0 INTRODUCTION

This chapter considers design factors and how they affect operations and maintenance (O & M) for sanitary sewer systems. Many of the topics discussed in this chapter also pertain to storm and combined sewer systems. However, the inclusion of specific design issues for storm and combined sewers is beyond the intent of this manual; this information can be found in several of the suggested readings listed at the end of this chapter.

Criteria used in a particular sewer design typically are established in the preliminary stages of the project. It is advisable to seek input from internal design staff, maintenance managers, superintendents, operators, geotechnical engineers, and construction personnel during the development of design and routing criteria. People in these positions know where field problems are likely to be encountered; thus, they can offer valuable information in the early stages of the project.

Key design considerations that affect O & M include:

- Line locations, grades, and depth;
- Route geography;
- Manhole spacing and size;
- Manhole channels;
- Influences of pumping stations;
- Odor control provisions;
- System instrumentation and controls;
- Induced turbulence (such as drop manholes);
- Sewer and manhole accessibility;
- Allowance for inflow and infiltration (I/I) capacity;
- Knowledge of system flow characteristics;
- Initial sewer flowrates compared to design flowrate;
- Degree of construction observation; and
- Proficiency of the sewer maintenance department.

Neither routine maintenance nor emergency service can be successful if maintenance personnel do not have appropriate access. One way to encourage the designer to include maintenance personnel is to require the maintenance manager's approval of maintenance on final design plans.

Ideally, collection systems should be designed to minimize total life-cycle costs; that is, the initial cost plus the present valve of O & M costs. This implies that a well-designed system can be easily and efficiently operated and maintained. However, some well-designed systems are not constructed as designed because of poor construction techniques, field changes, the absence of construction observation, and inadequate testing. Regardless of the reason, most collection systems will experience some deterioration and require general maintenance.

1.1 Regulatory and Environmental Requirements

1.1.1 *National Pollutant Discharge Elimination System Permits*

National Pollutant Discharge Elimination System (NPDES) permit language contains standard references to collection systems in four permit conditions: noncompliance reporting, proper operations and maintenance, duty to mitigate discharges, and prohibition and defense provisions. Designers should review these requirements with the collection system owner and consider whether state or federal clarification of the permit requirements is necessary. For instance, are the noncompliance reporting requirements clear? Are there exceptional circumstances where prohibition and defense provisions need to be expanded or more specific?

Entities that operate and maintain a collection system, whether or not they have an NPDES permit, are responsible for any discharge to waters of the United States from their sewers. Municipalities or permit holders should be sure that local ordinances are in place that clearly establish the responsibilities and controls expected from flow-contributing private laterals and private satellite collection systems that are served and do not have an NPDES permit.

1.1.1.1 Sanitary Sewer Overflows and Combined Sewer Overflows
The U.S. Environmental Protection Agency (U.S. EPA) and local state organizations have developed a much more active awareness of sanitary sewer overflows (SSOs) and combined sewer overflows (CSOs) in the last several years. This awareness has led to the creation of programs to not only eliminate overflows, but to ensure they do not occur or are drastically minimized.

1.1.1.2 Capacity, Management, Operations, and Maintenance
To help eliminate and control SSOs and CSOs, U.S. EPA developed a capacity, management, operations, and maintenance (CMOM) program designed to ensure collection

systems have effective programs in place. Moreover, it could provide a means of reporting SSOs and CSOs from collection systems. Although U.S. EPA has not officially taken on the role of regulating CMOM for collection systems, state and local agencies are serving in this capacity.

California has taken the CMOM program seriously, requiring all collection systems in the state to prepare sewer system management plans. These plans require the collection system operator to provide the following ten elements:

- Goals,
- Organization,
- Overflow emergency response plan,
- Fats, oils, and grease control program,
- Legal authority,
- Measures and activities,
- Design and construction standards,
- Capacity management,
- Monitoring, measurement, and program modifications, and
- Sewer system management plan audits.

If the agency feels any element is not applicable, it merely needs to address that in its document. The program is a monitoring and reporting tool that will not only track SSOs and CSOs, but ensure that collections systems are operated and maintained properly. For further information, please visit http://www.cwea.org/conferences/sso/Reg2Letter-SSMP0705.pdf and http://www.cwea.org/pdf/news/sso-wdr-factsheet.pdf.

1.1.2 System Conveyance Design Rates

Collection systems, or major hydrological sections of the system, should have an overall design rate that is accepted by state or federal regulatory agencies. Many communities develop this design rate (typically related to a storm event that might return in 5 or 10 years on average) during facility, sewer master, or watershed planning using historical and economic data. Design rates should include the planning period's wastewater flow projections and an I/I allowance that reflects existing or targeted seasonal and rain event flow variances. Establishing an overall conveyance design rate will provide the basis for new construction and rehabilitation design criteria and help avoid hydraulic bottlenecks or constrictions in the system that can lead to SSOs or CSOs.

1.1.2.1 Risk-Based Design Options
Recently there has been a trend to look at collection system design rates from a risk-based rather than storm-event based perspective. This approach allows a community to decide whether it is prudent to spend the dollars for a system able to control the design storm event now or to install a system that would handle the majority of the flows and be less expensive. This risk analysis compares the costs and risks associated with controlling or not controlling flows from a large future storm event for different scenarios. Further discussion may be found in the Water Environment Federation's *Wastewater System Capacity Sizing Using a Risk Management Approach* (2007).

1.1.2.2 Wet-Weather Facilities
Wet-weather facilities are designed and constructed to manage or control flows associated with wet-weather conditions. These facilities help protect the environment and public health. They may be designed to discharge under certain conditions upstream of a wastewater treatment plant. Any facility designed with an anticipated discharge must be permitted; however, wet-weather facilities may be designed for offline or inline storage with no discharge.

Wet-weather facilities may include one or a combination of processes, including screening, primary sedimentation, inline or offline storage basins, chlorination, or dechlorination (site-specific stream conditions may not require storage). For example, the facility will be designed to function according to site-specific, wet-weather conditions whereby the collection system wet-weather peak flows are diverted to wet-weather facilities for storage. If the diverted flowrate or volume is greater than what the facility can handle, excessive flow is discharged to a receiving stream after having gone through as much of the permitted process layout as possible. When wet-weather flows diminish, the stored wastewater is released back to the collection system.

Wet-weather facilities can help prevent problems associated with basement backups and reduce, through their containment functions, the frequency of SSOs. Wet-weather facilities must undergo a comprehensive evaluation and justification process, but can be the most reasonable, cost-effective alternative to larger conveyance systems that are unable to achieve adequate cleansing velocities during normal flow conditions.

1.1.3 Design Approvals
Conveyance system design approval procedures vary between states and localities. The designer should first discuss the design standards to be used with the owner of

System Design Considerations   103

the facilities. If all else fails, the designer should contact state, county, or other agencies to learn of any mandatory design standards or criteria that need to be addressed to receive regulatory approval.

1.1.4 Environmental Issues and Compliance

New construction and rehabilitation in collections systems typically can affect several different aspects of the environment:

- Endangered species,
- Air quality,
- Traffic,
- Waterways, vegetation, and paved areas, and
- Water quality.

Environmental issues should be investigated and cleared before design so that mitigation is understood and included in the design documents.

1.2 Permitting

Permit requirements also affect design. The designer should fully understand permits that will be required and the conditions of such permits before finalizing their design. Permits can range from local encroachment to U.S. Army Corps of Engineers 404 permits. It is a good idea to discuss permitting with the owner of the system because they may have a better understanding of requirements.

2.0 DESIGN GUIDELINES

2.1 Design Process Checklist

To aid the designer, a checklist should be used to ensure the necessary steps are followed:

- Project manager and are team assigned;
- Project intent clarified (new development, replacement, rehabilitation, etc.);
- Project schedule and budget established;
- Route selected;
- Alignment study completed for pipeline in selected route;
- Public, traffic, and environmental effects understood;

- Permit requirements considered;
- Survey conducted;
- Utility identified and potholing done;
- Geotechnical investigations completed;
- Hydraulics and pipe size determined;
- Pipe material selected;
- Manhole type and spacing requirements established;
- Pipe installation configuration and materials completed;
- Laterals and connections defined;
- Traffic control planned;
- Corrosion and odor control in place;
- Right-of-way needs and staging areas addressed;
- Special design features like tunnels, structures, and appurtenances incorporated;
- Input from O & M personnel, especially future access needs considered;
- Basis of design report established;
- Design submittals at 50%, 90%, 100%, etc. completed;
- Value engineering and constructability after 50% if required;
- Cost estimated;
- Specifications outlined;
- Bidding requirements clarified; and
- Construction supported.

2.2 Project Management

To have a successful project that meets the design intent, the project manager will have to help control schedule and budget and ensure that the final product satisfies the original requirements. This includes assigning proper personnel to assist with the aspects of the project and that tasks are accomplished on target. The project manager will also need to ensure that the quality of the documents is maintained and obtain required approvals. The manager will lead the remainder of the team through a planned process to complete the project. Tools used by the project manager might include scheduling software, budget tracking using a spreadsheet or database, presentation software, and communication (word processing) software. The manager will ensure that each component of the project is completed at its prescribed time and of the right quality.

2.3 Using Consultants

Consultants have been used successfully to handle project management as well as other aspects of design. Typically there is a qualification phase, a proposal phase, and a selection phase that may or may not include interviews with the team. Perhaps the most important aspect in the selection process is to select a firm and team that has completed recent, similar projects.

2.4 Routing Considerations

2.4.1 Assessing Aboveground and Belowground Interference

Routing considerations will be different for gravity versus pressure sewers, but both should emphasize O & M accessibility in their design criteria. Many factors affect the final selection of a route. Important considerations for aboveground systems include

- Paved versus earthen surfaces;
- Surface water flooding potential;
- Proximity to residential and commercial buildings;
- Construction problems such as limited equipment movement or stability risks of adjacent structure;
- Traffic or recreational disruption;
- Biological, aquatic, or historical disruption; and
- Number of land parcels traversed and easement requirements.

Important considerations for belowground systems include:

- Geology and soil conditions;
- Groundwater, river, or tidal influences;
- Other utility conflicts or influences (such as cathodic protection systems); and
- Depth of trench and safety requirements.

2.4.1.1 Surveys
It is important to ensure quality field data in conveyance design. It is imperative to obtain the best possible description of existing conditions, including existing utilities and surface features because it will be too late to make adjustments once construction begins. Carefully selecting the surveyor based on previous experience is just as important as selecting the designer. Ideally, they have worked together in the past. Location of utilities above ground may not be adequate and potholing may be necessary.

2.4.1.2 Potholing
Potholing refers to locating underground utilities by vacuum extraction or other means to determine depth and horizontal alignment. It is critical in cases where there are existing pressure pipelines such as water, gas, or explosive substances. However, fiber optic and communication lines also are critical; damage can affect an entire community. Adequate funds should be set aside to do a good job and locate as many utilities as needed. Funds spent in the design phase will save costs from changes during construction.

2.4.2 Geotechnical

Geotechnical surveys can reveal important information to the designer that will not only affect design and construction costs, but also will influence the life of the conveyance system. Deterioration of sewer and manhole materials can be accelerated if there are not adequate provisions for protection from corrosive soils. Surveys can reveal the extent of rock in an area and influence the feasibility of whether to install a gravity or pressure system. Surveys also can reveal soil expansion and stability conditions so that the designer can select materials and systems that will maintain water tightness while allowing for some lateral or longitudinal movement within the pipe and bedding zone. Typically geotechnical data should be obtained at approximately every 300 m (1000 ft) or more often depending on the anticipated changes in soil characteristics and makeup. For tunneled crossings, geotechnical data should be obtained at the proposed beginning, middle, and end locations. This information can help determine the type of tunnel to build and equipment needed to construct it. In an open-trench construction, the geotechnical data can provide recommendations for sheeting and shoring requirements.

2.4.3 Traffic Control

Depending on the final location of facilities, it may be prudent to provide guidance to the contractor for minimum traffic control features. Flagging may not be enough protection. Managers should consult with the local agency having jurisdiction of the roadway to determine the standards that they expect to be followed for traffic control during construction.

2.4.4 Permanent Access

Many I/I and maintenance-related problems occur because a sewer is inaccessible and cannot be routinely evaluated. Permanent access roads or gate assemblies on unimproved land should be considered a part of new sewer construction projects where possible. The cost for constructing these access roads is not likely to decrease in the future, and most sewer contractors will have suitable materials and equipment on hand to perform this task. However, construction of an access road across a wetland area should be evaluated carefully. The decision of whether to provide permanent access must be coordinated with permitting and other interested groups because wetlands will vary in physical character. Some wetland areas may require too much time, effort, and cost to enable access. The manager should look for environmentally sensitive, less-expensive, and less-controversial alternatives.

The designer should work closely with the sewer system owner and legal staff to discuss the required temporary and permanent easement widths. If permanent widths are too narrow for the sewer depth, repair costs are likely to be significantly greater in the future because of extra trench shoring and accessibility restrictions. This is particularly true as the area becomes more developed and easements encroaches occur.

The choice of initial routing location should allow for the trans-section of as few land parcels as possible. Designers and sewer system owners tend to underestimate the time needed to obtain easements. It is advisable to have all easements obtained (or at least critical easements) before awarding the contract or beginning construction. A process and schedule for obtaining easements is as important to the overall project as developing construction documents.

2.4.5 Public Effects

Because the public is affected by these projects, their needs must be taken into account. Otherwise politics will drive the solution rather than engineering. When a project will affect the public-such as because of changes in traffic flow and patterns-it is best to hold public meetings and seek input. It is also useful and beneficial to communicate with local law enforcement agencies when entering into highly disruptive traffic control situations.

2.4.6 Rights-of-Way and Easements

The designer needs to ensure proper access for construction and future maintenance by creating cross-sections of pipeline alignment and indicating several key items:

- How the trench will be excavated with backhoe clearances.
- How materials from the trench will be hauled away.
- How materials will be returned for bedding and backfill.
- Onsite storage and staging needs.
- Pipe delivery method and installation to the trench.
- Adequate clearances to accomplish traffic control.
- Future maintenance traffic needs and access.
- Need for security.

Permanent easements must provide adequate room for future excavation in case of replacement or rehabilitation. Vehicles and equipment will need to be able to access manholes easily for inspection and cleaning. Typically easements are centered on the installed pipeline; accessibility, however, should dictate location. It is not unusual to need approximately 9 to 14 m (30 to 45 ft) for pipeline access.

Temporary easements range in size depending on the project and can be as much as 15 m (50 ft) wider on one side or both sides of permanent easements. For a 1829-m (72-in) pipe, 37 m (120 ft) total easement would not be unreasonable. It is also important to recognize that temporary easements may be needed to provide staging areas for equipment or materials or for supply storage.

Easements are rights given from one owner to another. If the pipeline will traverse a public right-of-way or street, easements are easier to obtain. Private property raises other considerations. If the owner will not grant the easement, consideration must be given to other recourse such as eminent domain (condemnation). All these issues must be taken into account during design.

2.5 Hydraulic Analysis Using Modeling

For most design projects, the sizing data provided in the paragraphs below will be adequate for determination of pipe size, slope, and velocity. However, in some cases, an entire system master plan may be required, if, for instance, determining flow is more complicated. In these cases, the designer can use hydraulic models for sewer system analysis. The designer will need to work with the system owner to determine how the hydraulics should be handled. A discussion of the use of hydraulic computer models for system design and capacity analysis is provided in Chapter 4.

2.6 Pumping Stations

2.6.1 Package Stations

Sewer systems cannot always flow by gravity, and, in many small systems, the need for pumping stations can be handled by purchasing package lift stations. These can range from submersible systems that can be installed in standard concrete manholes to more elaborate self-contained systems with full-service electrical gear. The wet well (location where wastewater enters and is stored) typically is the only major structural component. There are, however, package stations with both wet and dry wells (dry wells are used to keep pumps dry and out of the wastewater atmosphere). Regardless of station type, it is good practice to provide redundancy and a backup pump in case of duty pump failure. Equipment manufacturers can provide design references for this type of pumping station.

2.6.2 Other Types of Stations

Although this document cannot describe the full design features for pumping stations, larger pumping stations should contain several basic components:

- Wet well.
  - With submersible pumps only (screening equipment not provided) and
  - With dry pit pumps (screens will be required ahead of wet well entrance).
- Drywell (if not submersible pumps).
- Screening or grinding equipment ahead of wet well entrance.
- Pumps (redundancy should be provided by having at least one duty and one standby pump for each size pump provided).
  - Submersible in wet well and
  - Dry pit centrifugal or other.
- Pump controls.
- Electrical supply.
- Supervisory control and data acquisition or other operating equipment.

Pumping station design is covered in more detail in *Design of Wastewater and Stormwater Pumping Stations* (WEF, 1993).

2.7 Pipelines and Joint Materials

2.7.1 Gravity Pipelines

2.7.1.1 Sizing Considerations

Gravity pipeline sizing should result from applying design criteria to system conveyance-planning flow projections. In addition, there are many reference texts and manuals that provide concepts for sizing a gravity sewer, such as *Gravity Sanitary Sewer Design and Construction* (ASCE and WEF, 2007). The most common method of gravity pipeline design is the use of the Manning equation where capacity or flowrate is based on cross-sectional area, hydraulic radius, slope, and roughness coefficient known as the "*n*" factor. Typically Manning's "*n*" will vary, but 0.013 is used widely. However, selection of the pipe diameter is increasingly being influenced by O & M considerations, such as scouring velocities or providing wet-weather, inline storage.

Some communities have adopted sewer criteria that establish a minimum velocity of 0.75 to 0.90 m/s (2.5 to 3.0 ft/sec) rather than 0.61 m/s (2 ft/sec) at half- or full-pipe flow conditions. Other communities have oversized their interceptors to provide inline storage during wet weather and incorporated O & M activities to induce cleansing velocities. Camp's equation, provided below and in the reference cited above, can also be used to determine cleansing velocity in sewers:

$$V = [(8B/f)g(s - 1)D_g]^{1/2}$$

Where

V = velocity (m/s, ft/s),
B = constant ranging from 0.4 to 0.8 (dimensionless),
f = Darcy–Weisbach friction factor (dimensionless),
g = gravitational acceleration (9.8 m/s², 32.2 ft/s²),
s = specific gravity (dimensionless), and
D_g = diameter of particle (m, ft).

or

$$V = (1.486/n) R^{(1/6)} [B(s - 1)D_g]^{1/2}$$

Where

V = velocity (m/s, ft/s),
n = Manning coefficient of roughness,
R = hydraulic radius (m, ft),
B = constant ranging from 0.4 to 0.8 (dimensionless),
s = specific gravity (dimensionless), and
D_g = diameter of particle (m, ft).

Increasingly available and affordable hydraulic modeling programs are informing decisions regarding oversizing and pipe and joint sizing. Risk-based capacity design is covered in a *Wastewater System Capacity Sizing Using a Risk Management Approach* (WEF, 2007).

2.7.1.2 Materials
Proper sanitary sewer design includes careful evaluation of construction materials. Future maintenance and rehabilitation costs can be reduced by addressing factors that affect the life expectancy and use of the sewer. Wastewater characteristics, site conditions, and sewer construction methods affect material selection. Factors to consider when evaluating a gravity piping system include

- Life expectancy and performance history;
- Availability of sizes required;
- Backfill and bedding material requirements;
- Availability and ease of installation of fittings or connections;
- Type of joint (fluid and pressure tightness and ease of assembly);
- Resistance to scour or abrasion;
- Resistance to acids, alkalis, gases, solvents, and other industrial wastes;
- Ease of handling and installation;
- Physical strength;
- Flow characteristics (friction coefficient);
- Cost of materials, handling, and installation; and
- Manufacturer testing and quality control program.

No single material is optimal for all conditions; selection typically is made based on the particular application under consideration. Different grades of materials may be selected for portions of a single project, but this practice typically is kept to a minimum unless a significant cost savings would result. The differential in the costs of various types of pipe tends to be a small part of the total project cost but may be worth considering as a more expensive pipe and manhole may reduce future O & M costs. *Gravity Sanitary Sewer Design and Construction* (ASCE and WEF, 2007) describes the more typically used materials and their advantages and disadvantages.

The basic pipe materials commercially available for gravity sewer construction and given in the references are:

- Ductile iron,
- Reinforced concrete,
- Polyvinyl chloride (PVC),
- High-density polyethylene (HDPE),
- Fiber glass, and
- Vitrified clay.

Ductile iron and concrete pipes typically have interior linings or exterior coatings to extend their life expectancy because of the aggressive and corrosive sewer environments to which they are subjected. Ductile iron can also include a polyethylene bag or wrap to prevent external corrosion from aggressive soils.

2.7.1.3 Backfill and Bedding

For the purposes of this discussion, backfill will include the zone of material around the pipe in a trench installation. The area immediately above, below, and around the pipe will be called the pipe bedding. In practice, initial, base bedding is placed, the pipe is installed on that surface, and then final bedding is extended above the pipe to adequately protect it. Both the pipe backfill zone and the bedding area serve an important function in the pipe's ability to support the resultant load forces. The materials that can be used in the pipe zone and bedding area vary as widely as the soils in which they are buried. Therefore, it is important for the designer to coordinate the pipe's inherent strength characteristics with the type of backfill and bedding material that will be used during construction.

Attention to backfill and bedding requirements during design and construction can eliminate many typical O & M problems. Improper pipe backfill and bedding design or a lack of uniform bedding during construction can lead to pipe structural failures, differential joint settlement, pipeline sags, and admission of groundwater, soil, and roots. In addition to the pipe manufacturer's design guidance literature, there are several good references the designer can review for a more detailed discussion of backfill and bedding design practices, such as *Gravity Sanitary Sewer Design and Construction* (ASCE and WEF, 2007).

Designers should consider whether to include an allowance for future earthen loads when selecting a pipe and backfill combination. For instance, pipelines routed along undeveloped urban streams or ditches may experience in their design life several meters (feet) of fill material greater than the original earthen depth. This often happens when a developer wants to raise the foundation elevation of proposed new buildings above the flood elevation.

Deviation from designed trench or embankment widths during construction may result in an overload to the pipe. This is particularly true when a sewer pipe approaches and leaves a manhole. Designers should be aware that if a limited trench width is required for the backfill material specified, and if it is less than the width that is required to place a manhole in the trench, the potential exists for an overloaded pipe. Also, inadequate backfill compaction could cause differential settlement, overdeflection, or shearing, which can lead to pipe failure. An evaluation of the probability of differential pipe settlement versus structure settlement is part of the design phase.

Granular fines can be scoured from a compacted backfill if the terrain grades are steep. The designer should consider installing a solid encasement around the pipe or a water barrier such as clay, bentonite, or other impervious material in the trench to impede subterranean flow along the pipe. Fines can migrate from the native soil in the trench walls into the granular backfill as well. This can cause a loss of lateral support and additional load-bearing forces on the pipe. Flexible and semirigid pipe materials will be affected by the loss of lateral backfill support and possibly experience loads greater than what the designer had anticipated. One method to reduce fines migration is to line or wrap the trench bedding and backfill zones with a geotextile material (sometimes called a "burrito wrap"). This method also is also when native soils in the trench are soft organic clays or peat that might allow the granular material to move laterally when placed under load.

The O & M problems and pipe failures previously mentioned also can occur when the pipe diameter becomes sufficiently large that the installer has difficulty placing and compacting the backfill under the haunches of the pipe as required. This problem is more common for pipe diameters greater than 61 cm (24 in.). A flowable fill (grout) or controlled density fill is an alternative; however, other issues such as the potential to float the pipe and chemical attack to metal pipes and appurtenances become considerations that also must be addressed.

Insufficient cover over a pipe can cause pipeline problems similar to those previously discussed. Operations and maintenance problems may develop faster with shallow cover because the failed or defective pipe is more susceptible to roots- or rain-induced I/I. In colder regions of the country, pipes should be installed below the frost line.

2.7.1.4 Grades
The composition of domestic wastewater has changed as the public's needs and practices have changed. Many convenience wastes, such as non-returnable or disposable materials, find their way into sanitary sewers. Proper sewer grades help keep these wastes from settling.

Too often, sanitary sewer designs include long lengths of line laid at minimum grades, typically accepted as grades that produce 0.61-m/s (2-ft/sec) flows at half- and full-pipe diameters. A somewhat steeper grade that produces velocities of 0.76 to 0.91 m/s (2.5 and 3 ft/sec) typically is preferable to reduce deposition and hydrogen sulfide generation.

On steep grades, care must be given to the selection of pipe materials to prevent pipe wall erosion when velocities are greater than 1.8 m/s (6 ft/sec). At these velocities, the designer should consider using a drop manhole concept to reduce the flow velocities. Also, special attention to anchoring to prevent the pipe from creeping downhill should be installed. Anchors or retards should be coordinated with the appropriate backfill material. For instance, as previously mentioned, granular backfill in steep trenches may shift, causing the pipe to pull apart, particularly if percolating water velocities scour the backfill fines.

Sanitary sewers are designed for future peak flow. For this reason, problems may be encountered in minimum-grade sewers during the early years of their design life. Low initial flows that do not approach design capacity can create maintenance problems and may require frequent line flushing and cleaning.

2.7.1.5 Inflow and Infiltration Allowance
Most sewer systems do not stay as watertight as they were on the day of their installation and performance test. After a new sewer system is installed, it eventually develops defects or avenues for extraneous water to enter. Often, the older the system is, the greater the I/I; however, newer systems can sometimes exhibit as much or more I/I, depending on the quality of construction and local conditions.

Design allowances for I/I are best developed for new pipe and collection systems from flow monitoring data collected on the existing sewer system or by benchmarking against other similar sewer systems. Allowances for I/I should be coordinated with design standards that result from system conveyance planning design rates previously discussed.

2.7.2 Pressure Pipelines

2.7.2.1 Types of Pressure Systems

Pressure systems can be either positive or negative. Most systems are positive-pressure, whereby a pump transfers energy to lift the wastewater to the desired elevation at a specified rate. The most common positive-pressure system results when a pumping system receives wastewater from a gravity sewer or other pumping systems. Pipeline diameters for these systems typically are 10 cm (4 in.) or greater. System pressures typically range from a few meters to approximately 60 m (200 ft) of head, with the norm being between 15 and 30 m (50 and 100 ft) of head.

Another form of positive-pressure system is a low-pressure pumping system wherein a pump receives flow from a single facility or a small group of residential or commercial facilities. The type of low-pressure pump used is influenced by whether the pump receives the wastewater directly into the sump pump or from an existing settling compartment such as a septic tank. The designer has the basic option of using a centrifugal or positive-displacement pump. Residential systems that use grinder pumps at residence are also common in many parts of the country.

The pressures produced typically are less than 46 m (150 ft), and the discharge flows to a common pumping station wet well or a gravity system. Pipeline diameters from low-pressure pumps are typically 4 cm (1.5 in.) and pump to a 5- to 10-cm (2- to 4-in.) diam header pipe. These systems can be used where geographical or topographical features would make a gravity sewer system cost-prohibitive, such as in areas along lakes, bayous, or other waterfronts that either have high groundwater or variable terrain elevations. Low-pressure systems also are used in areas of rocky terrain or where housing is sparse.

Infiltration and inflow into both traditional and low-pressure pumping systems is relatively minor compared to gravity sewer systems. Essentially, the only component of the pressure system that is subject to I/I entry is the sump or pump wet well.

A negative-pressure system (also referred to as vacuum sewer) produces a negative pressure within the pipeline between the residential or commercial facility and the pump. The wastewater is drawn to the pumping station where it discharges to a gravity sewer or another pumping system. A negative-pressure system typically has a pipeline diameter of 5 to 20 cm (2 to 8 in.). Such systems have the same applications as the low-pressure systems previously discussed. Negative-pressure systems also are subject to minor I/I because of the potential for leaks. For additional detail on systems such as pressure, vacuum, and small-diameter gravity sewers, see *Alternative Sewer Systems* (WEF, 2008).

2.7.2.2 Sizing Considerations

Most piping systems function as a single-barrel pipeline; however, a dual or parallel pipeline is not uncommon where service redundancy is extremely important or where clearance conflicts might prevent a single-barrel pipeline from being installed. In either situation, sufficient internal minimum velocities of 0.61 m/s (2 ft/sec) and, preferably, 0.91 m/s (3 ft/sec) must be reached to resuspend organic and small, sand-sized solids that enter the pipeline. As with gravity sewers, the early years of a new installation may see flows that are at the low end of the design range. Therefore, the designer should evaluate whether minimum velocities will be reached daily. If not, then it may be necessary to develop a weekly or more frequent O & M plan to create enough flow to flush the pipeline. The extent of the flushing plan will depend on length of the pipeline. As an alternative, it may be necessary to include a pipeline pig launcher and retrieval appurtenances in the system design. A pig is a plug-type insert that in the pipeline that uses either hydraulic force from increased fluid velocities forced between the pig and the pipe wall or hydraulic pressure to mechanically scrape or dislodge material from the pipe.

Maximum velocities for wastewater positive-pressure systems should not exceed 3 m/s (10 ft/sec) for normal, sustained flows. Wastewater conveys sand and grit that accelerates the inside deterioration or erosion of pipe coatings or wall thickness at bends or valves. Velocities from 0.9 to 1.8 m/s (3 to 6 ft/sec) are more desirable for sustained flows. The ability to stay within these ranges will depend on the flow range that the pipeline must convey.

2.7.2.3 Materials

Typical pipe materials commercially available for pressure sewer construction include:

- Ductile iron,
- Prestressed concrete cylinder pipe,
- Polyvinyl chloride,
- High-density polyethylene, and
- Steel.

The same evaluation considerations apply to pressure pipe materials as previously discussed in the selection for gravity pipe materials. However, there are additional O & M considerations when evaluating pressure pipes. Pressure pipes tend to be installed at shallower depths than gravity sewer pipes. In colder regions of the country, pressure pipes should be installed below the frost line. Therefore, native soil character, surface contaminants that percolate through the soil, and stray currents from other protected structures can influence the exterior corrosion rate of the pressure pipe. Interior corrosion can be more aggressive in pressure pipes at high profile points and at the discharge to the atmosphere.

Pressure pipes, particularly nonmetallic, should be installed with a system for recovering the location of the installed pipe. This can be done with accurate record drawings or by burying a wire, tape, or electric signal coupon above the pipe.

2.7.2.4 Backfill
Backfill has the same important structural support function as discussed for gravity sewer pipe. Soil conditions can vary widely in a project, and it is typical to use local, excavated soil as backfill around the pipe. Geotechnical engineers should assist in closely evaluating soil characteristics, which influence the performance of thrust restraint for pressure systems.

Earth loads typically are not as great an external load because of the shallow depths sometimes involved. However, live loads can become significant. Designers should thoroughly review the pipe manufacturer's backfill design suggestions when designing the pipe and bedding combination.

2.7.2.5 Restrained Joints
The design of thrust restraint for pressurized, buried pipelines with push-on-type joints is important because of the damage that can result from joint separation. Pipe systems that have all joints restrained, either as furnished by the pipe manufacturer or from a welded joint, may not require external thrust restraint.

Thrust occurs when pipeline size or direction changes. Thrust forces can be hydrostatic, hydrodynamic, or thermal in origin; designers are responsible for evaluating each type of force. The designer should review the pipe manufacturer's recommendations for restrained joints and determine if application of the restrained pipe length and method is appropriate for the proposed site conditions.

Resistance to thrust forces can be provided by concrete thrust blocking, concrete thrust collars, or anchorage to a structure, or, indirectly, by transmitting the forces to the adjacent straight pipe sections using joints designed to resist shear and separation. The forces in the adjacent pipe are then transmitted to the soil by a combination of friction and lateral pressure.

Concrete thrust blocks depend on the allowable lateral soil pressure to develop the necessary resistive force. The allowable lateral soil pressure for local conditions must be determined during design and properly observed during construction. In adequate soils and for smaller diameter pipe, thrust blocks typically are the most economical method of resisting thrust. However, sometimes soils are unable to resist lateral force or dead weight force of the block. Also, there are times when space limitations or the probability of future excavations will dictate the need for an alternative restraint method.

Designers should also check for corrosion requirements to determine what protection is necessary for the thrust-restraint system.

2.7.2.6 Blowoff and Air Valves

Air valves should be an integral part of the pressure pipeline system. Air that enters or is entrapped in the pipe must be released. Because flowing water is subject to changing pressures and velocities, air is continually coming out of solution when the pipeline is in service. Unless air valves are properly located, this free air will accumulate at high points within the system. Serious flow restrictions or even blockage can result from air pockets in the pipe, which can contain corrosive vapors and should be vented to reduce interior deterioration.

Surge-pressure forces will be generated in the pipeline during situations such as pump startup, rapid shutdown, or sudden valve closure. The forces created can be substantial and need to be evaluated to determine their influence on the design. Whether surge suppression is accomplished through a surge-relief tank or air release valves, provisions for their regular maintenance should be established.

Valve maintenance is especially important on wastewater pipelines. Corrosion-resistant materials for the entire assembly often are overlooked. Use of corrosion-resistive materials should be considered throughout the system from the corporation stop through the shutoff valve and air valve internal parts. Otherwise, when a leak develops, maintenance personnel might not be able to close the shutoff valve or the corporation stop.

The designer needs to ensure correct sizing and selection of the air valve. Technical bulletins published by the manufacturer or included in reference materials will have recommendations for making the correct selection. These materials also will include recommendations for equipment and procedures for periodic valve cleaning. The designer should take into consideration these maintenance requirements and provide adequate access space for a maintenance worker to repair or remove a valve.

This could include providing proper vault drainage to prevent contamination of adjacent surface water or groundwater and including provisions for odor control if the valves are close to residential or commercial establishments.

Similarly, sediment and debris can accumulate in low points; methods to remove these items need to be provided during design. Consideration should be given to avoid low points; when needed, blowoff valves should be provided at low points.

2.7.3 Odor and Corrosion Control

Without proper consideration of odor and corrosion control, the wastewater conveyance O & M staff will encounter continuous maintenance and repair problems from a deteriorating piping system.

Odor control may be accomplished by many different methods including reducing turbulence in the piping system, using activated carbon on air valves or for pumping stations, or adding chemicals or biofilters. The designer should understand potential odor issues and take measures to control odor as required.

This section will not discuss corrosion control methods in detail but will outline important considerations in the selection of steel, iron, and concrete pipe materials for wastewater conveyance applications because these are the most susceptible to corrosion. A corrosion engineer or manufacturer representative should be consulted for specific corrosion protection controls. In addition, the designer should not overlook metallic components (such as fittings, bolts, bands, and valves) that are used with PVC, fiber glass, and HDPE conveyance systems.

Corrosion is a result of the reaction between the pipe material and its environment. Corrosion can occur on the interior of the pipe as a result of reactions with wastewater. The exterior can corrode because of reaction with the soil. Designers often fail to anticipate the corrosion potential of electrical wiring, switches, and fixtures. The factors that must be considered when evaluating materials are determined after field testing and evaluation of the environmental conditions.

Bonded pipe coating and linings help control corrosion. They isolate the pipeline from its environment. However, coatings or linings are not perfect, and additional techniques may be required to control corrosion. Portland cement concrete or mortar can also be used to protect steel in non-detrimental environments. This is the basis for protection of the steel components of concrete pressure pipe and for mortar-coated and lined steel and iron pipe.

Polyethylene encasement is an unbonded coating traditionally used to protect the exterior of ductile iron pipe. This system uses polyethylene sheets that are loosely wrapped around the pipe and fittings. This barrier reduces the exterior corrosion by reducing exposure to oxygen, corrosive species, and stray currents. Polyethylene encasement may provide economical control from external corrosion in certain environments. However, it is important that the polyethylene encasement system be installed correctly to provide the necessary protection.

Cathodic protection is useful for mitigating external corrosion of both metallic pipelines and concrete pressure pipe. Cathodic protection can be an external source of direct current voltage applied to the pipe so that the pipe becomes lower in electrochemical potential than the soil around it. Thus, the pipe becomes the cathode rather than an anode. Sacrificial anodes such as magnesium or zinc rods can also be attached to the metal pipeline and buried in the soil around the pipe. The magnesium or zinc rods corrode sooner than the metal pipe because the electrochemical force developed between the metals causes the anode to lose the metal ions before the pipe does.

Application of cathodic protection to a coated metallic pipeline supplements the protection already afforded by coating systems. Cathodic protection systems can be quite complex and should be designed by corrosion specialists trained in assessing when and how to implement this method of control. Cathodically protected systems require regular O & M to read and monitor the status of the protection system.

2.7.3.1 Sulfide Control Design Considerations

The following guidelines have been prepared to help minimize the production of dissolved sulfide and the release of hydrogen sulfide gas into wastewater collection systems.

Sulfide control guidelines for gravity sewers are as follows.

- Sufficient velocities should be maintained in gravity sewers to minimize the production of dissolved sulfide as described under design.

- Excessive velocities that would cause turbulence of the wastewater and stripping of hydrogen sulfide gas should be avoided. Flow velocities should not exceed limits previously discussed. The sulfide concentration in the wastewater to be conveyed by the gravity sewer should be considered before designing for velocities that could cause the stripping of hydrogen sulfide gas.

- Turbulence at points of intercepted flow (manholes or wyes) should be minimized through the use of smooth transitions. Hydraulic drops and outfalls should be avoided. The use of drop manholes should be minimized. Optional methods such as submerged-outlet drop pipes or contained drops should be considered for drops of up to 6 m (20 ft) to contain and minimize turbulence.

- Combining wastewater flows at a confluence angle greater than 90 deg should be avoided. Confluence angles of 45 deg or less have been found to reduce sewer pressurization, turbulence, and corrosion.
- It is recommended that sewers not be installed with opposing flows (confluence angle of 180 deg).
- The use of inverted siphons should be avoided whenever possible. Inverted siphons should be provided with adequate air jumper capabilities and positive corrosion protection measures such as use of liners or inert materials.
- It is recommended that diameter changes only occur at manholes that have been properly designed to allow for change in flow characteristics.

Sulfide control guidelines for pumping stations and force mains are as follows.

- Pumping stations and force mains should be designed to minimize hydraulic detention time within the pipeline, thereby reducing dissolved sulfide production.
- The discharge of force mains into the wet well should be designed to minimize turbulence and the stripping of hydrogen sulfide gas. Wherever possible, discharges should be submerged to absorb hydraulic energy and minimize turbulence.
- Gravity sewer entrances into wet wells should be designed to minimize free fall and turbulence. Pumping stations should be designed so that free air movement through the sewer is not impeded by wet well water levels except at flows greater than peak dry weather flow.
- The number of air release points (placed at high points) for force mains should be minimized through grade and slope control. Drainage for the force main should be provided if possible. To avoid localized corrosion, partially full force-main conditions should be avoided at any point in the pipe. Force mains should be designed with a flat or continuously rising profile whenever possible. If possible, force mains should not be designed with a downward or negative slope. The idea is to control grade as much as possible and place air release valves as needed.

- Dual-force-main piping should be considered. This will help maintain minimum velocities, provide for future capacity, and minimize dissolved sulfide generation (if force mains can be drained).
- The pumping station could be designed so that pumps clean the wet well. Self-cleaning wet wells may be advantageous for wastewater.
- Where the force-main profile cannot be controlled and air releases must be used, the use of variable frequency drives on pump motors should be considered to maintain flow conditions and avoid excessive operation of the valves.

2.7.3.2 Corrosion Protection Design Considerations

The following guidelines have been prepared to reduce corrosion potential in wastewater collection and treatment systems caused by exposure to hydrogen sulfide gas. In general, hydrogen sulfide gas concentrations less than 2.0 ppm are not considered corrosive. Hydrogen sulfide gas concentrations between 2.0 and 5.0 ppm are considered potentially corrosive, and concentrations greater than 5.0 ppm average justify protection against corrosion.

Corrosion protection guidelines for gravity sewers are as follows:

- Pipe material and pipe liners used for wastewater collection system construction should be unaffected by contact with sulfuric acid solutions as low as pH 0.5 (approximately 7% sulfuric acid by weight).
- Cementitious concrete products of any type (calcium silicate, calcium aluminate) should not be used for wastewater conveyance wastewater without an appropriately designed corrosion protection system (liner).

Particular attention should be given to corrosion prevention of existing, unprotected concrete gravity sewer pipes and manholes in areas of increased turbulence (intercepted flows, 90-deg bends, slope changes, drop manholes, and force-main discharge locations) for a distance of 300 diameters downstream and 50 diameters upstream.

Corrosion protection guidelines for pumping stations and force mains are as follows:

- Pumping station wet wells should be constructed of or lined with inert materials where hydrogen sulfide is anticipated. Plastic liners provide a longer life than applied coatings. Exposed, unprotected concrete of any kind should not be used in pumping station wet wells or other areas exposed to average concentrations of hydrogen sulfide gas greater than 5.0 ppm.

- Piping used inside pumping station wet wells or areas of treatment plants exposed to significant concentrations of hydrogen sulfide should be constructed of or lined with inert materials. The recommended metal is 316 stainless steel. Coated ductile iron is also not advisable.
- Force mains should be constructed of inert pipe materials or factory-applied polyethylene-lined ductile iron pipe. Prestressed concrete cylinder pipe is not recommended for wastewater force-main applications.
- Particular attention should be given to protection of structures downstream of turbulent force-main discharge areas. It is even better to avoid designs that cause turbulence.
- Force-main discharge manholes should be constructed from inert materials or protected with material such as a PVC liner or epoxy coating.

2.8 Rehabilitation

Rehabilitation maintains the structural and operational integrity of the sewer system. The methods available to evaluate and rehabilitate the sewer system have become more sophisticated with the advancement of robotics and improved camera systems. *Existing Sewer Evaluation and Rehabilitation* discusses this subject in more detail (WEF and ASCE, 2008).

Both gravity and pressure pipelines will require repairs or replacement during its lifespan. Many of the factors that create the need to repair or replace pipe have been discussed. Typically, the primary reason for repairing or replacing conveyance system pipe is related to structural or I/I problems.

2.8.1 Condition Assessment for Design

The most critical part of any rehabilitation program is to understand the pipe condition and need for repair or rehabilitation. Sewer system condition assessment technologies such as closed-circuit television (CCTV) and smoke testing are discussed in more detail in Chapter 4 of this manual. Many improvements have been made over the last 10 years in condition assessment techniques and methodology. Regardless of the method used, it is best when the designer and the system owner speak the same language.

2.8.2 Sewer System Assessment Protocols

The National Association of Sewer Service Companies developed sewer system condition assessment protocols called the Pipeline Assessment and Certification Program and the Manhole Assessment and Certification Program. These programs were developed to promote the use of common standards within the sewer industry including procedures, a coding system, and a consistent method of assigning severity codes to identify, quantify, and prioritize defects in an existing collection system. For further information on this program, please check http://www.NASSCO.org.

2.8.3 Structural Repairs

The need to make structural repairs typically becomes evident either through a sudden pipe failure or through field inspections such as internal CCTV inspections or pipe-wall coupon sampling. In each case, an exact cause of the problem needs to be determined so that the corrective action is effective. Otherwise, the corrective action may have to be significantly overdesigned to account for this imprecise assessment and all of the possible or speculative causes of the problem.

Structural repairs typically consist of replacing the section of piping system with the same material. It would be advisable to use a granular pipe-bedding material to provide stable, uniform support.

The designer also should consider using the pipe manufacturer's recommended replacement couplings and suggestions. If a section of pipe is replaced, the internal pipe diameter of the spliced-in pipe should match the internal diameter of the adjoining sections. Otherwise, the joint between the old and new pipe could become a point where solids and debris become lodged.

2.8.4 Technologies for Design

Repairs for both pressure and gravity pipelines also can made to the exterior using clamps or sleeves or to the interior using a liner either for long stretches or for several meters. For longer pipe segments, rehabilitation systems include slip-lining, swaged and rolled-down pipes, segmented linings, cured-in-place (CIPP) systems, and fold-and-form pipe (FFP) systems. The systems have become increasingly competitive and beneficial because of their less-disruptive trenchless technology features. Both CIPP and FFP systems have application for point structural repairs as well.

A critical principle to consider in structural design is that the load the liner must support is determined by the degree of degradation of the original pipe. Design conditions for liner structural design typically fall into two primary categories: partially deteriorated and fully deteriorated. There are many materials and systems for both external and internal materials that are commercially available. These materials and systems may be advertised in national trade organization publications.

Pipe bursting or pipe eating are other forms of pipe replacement. These systems break the original host pipe, and a new HDPE pipe is pulled into its place. Pipe upsizing of one to two diameters is possible under certain conditions. Both pipe bursting and liner systems require that individual service connections be reestablished either by using robotics inside of pipe or by excavating to expose and reconnect the service connection.

Please refer to *Existing Sewer Evaluation and Rehabilitation* for more data on pipeline rehabilitation means and methods (WEF and ASCE, 2008).

2.8.5 Inflow and Infiltration

Inflow and infiltration is a problem associated with gravity piping systems. The term refers to extraneous flow that enters underground sewer pipes and imposes an undesirable hydraulic load on the sewer system. Infiltration sources include groundwater infiltration, which may be prevalent year-round, and rainfall-derived infiltration and inflow, which is present during and shortly after rain events. Inflow enters the sewers via direct connections to the sewers such as through manhole covers, surface or yard drains, roof downspouts, sump pumps, or service connection cleanouts with missing caps. Stormwater exiting the storm sewer system through damaged pipes, non-water-tight joints, and unsealed lift holes may follow nearby sanitary sewer trenches and enter through damaged pipe and non-water-tight joints.

Rehabilitation approaches to correct infiltration may be different than those to correct structurally damaged pipe. Television inspection of the sewer can help define the extent of structural rehabilitation needed. Groundwater migration, however, will influence the required rehabilitation to control infiltration. Therefore, infiltration rehabilitation work may include sealing or making each sewer and service connection joint watertight the entire length of the sewer within a groundwater zone of influence. Rehabilitation might include chemically grouting each joint (or at least pressure testing each joint) or relining the original sewer with one of the many commercially available lining materials discussed above.

2.9 Manholes

Manholes are installed in sewers to provide access for routine or emergency maintenance. They are sometimes rendered inaccessible because of location, negligence (for example, covered by pavement), or design deficiencies. Locating sewer lines in street or alley rights-of-way is preferable to the use of easements, where obstructions such as utility poles, meters, junction boxes, fences, retaining walls, trees, shrubs, or utility buildings can restrict access. It is important to avoid placing manholes and sewer lines in the path of direct, continuous traffic; it is best to locate these items outside wheel areas.

Manholes typically are placed at each change in direction so that sewers are straight between manholes as a result of several considerations: straight sewer layout and construction is easier; accurate record drawings have often not been kept to allow relocation of curved sewers; and pulling CCTV and cleaning equipment in curved sewers is more prone to operational problems and pipe damage. Curved sanitary sewers are more prevalent in older systems even though a properly designed curved sewer will convey wastewater just as effectively as straight sewers and can cost less. Some communities, however, have adopted the practice of installing a manhole immediately upstream, downstream, or at both ends of the curved portion of the sewer if additional construction work is done in the vicinity of a curved sanitary sewer.

Ideally, manholes are large enough to provide space for a person to work with cleaning equipment, a shelf wide enough to provide safe footing, and support for a ladder. Manhole walls can be sealed from groundwater infiltration by using mortared joints, adequate flexible sealer, or rubber O-ring gaskets for precast-section manholes. It is reasonable to consider a combination of joint-sealing methods because concrete casting tolerances may not produce tight-fitting tongue-and-groove precast joints.

Manholes typically have an inside diameter of at least 1.2 m (4 ft). Larger pipelines need correspondingly larger manholes to provide an adequate shelf on which to work and a wall surface sufficient for a solid, watertight connection. Eccentric manhole shafts (i.e., shaft is offset to be aligned vertically with manhole wall) are easier and less dangerous to enter. To reduce problems with some maintenance equipment, the straight side is sometimes over the outlet pipe. Many agencies require construction of concentric manholes (i.e., entrance shaft is in center of manhole), which facilitates the insertion of cleaning equipment in multiple directions regardless of orientation. Because of safety considerations, some agencies have eliminated manhole steps and now use ladders or hoists.

2.9.1 Spacing

Manholes typically are spaced at intervals of not more than 120 m (400 ft) for small sanitary sewers. Spacing often is specified by local regulation. With the advent of modern cleaning equipment, some agencies have considered lengthening this maximum interval. Although some new cleaning equipment is able to clean up to approximately 300 m (1000 ft), other maintenance activities, such as bypassing a section or certain rehabilitation activities, can be difficult or impossible at that distance. Where there are truck-access limitations (such as in a backyard or parkway), the distance from the access point to the manhole needs to be considered in determining the manhole spacing. Intervals of 150 to 300 m (500 to 1000 ft) are typical for large sanitary sewers of 0.75 m (30 in.) diameter or greater.

A manhole also is necessary at each change of grade, invert elevation, or pipe size; at each intersection of two or more sanitary sewers; at the terminal end of each sewer line; and at major service connections (such as industrial facilities and schools). Manholes should not be placed at busy street intersections.

The practice of replacing the last manhole with a lamphole or vertical riser is not recommended. When sanitary sewers must be located along a rear property line, the sewer should be extended to a street right-of-way or an access easement, even if this requires construction of an additional dry sewer. This ensures access for the maintenance crews.

2.9.2 Materials

2.9.2.1 New Construction

Most new manholes are made of precast concrete. Other materials such as fiberglass and HDPE also are used in environments that are detrimental to concrete. Some communities require coal tar epoxy, elastomeric epoxy, or PVC liners in the manholes that are to be located on interceptors greater than a specified diameter. The walls of new manholes typically are heavily pitted and need to be filled with a cement layer before application of an epoxy liner. Otherwise, corrosive gases will penetrate existing pinholes, corrode the area behind the liner, and eventually delaminate the liner. Precast concrete manholes that are located in the upper reaches of collection systems or where hydrogen sulfide generation is minimal will provide years of good service.

2.9.2.2 Rehabilitation

Manhole rehabilitation typically begins with investigations to help identify I/I problems in the collection system. Many communities prefer to rehabilitate not only their I/I defects but also their structural defects under the same effort. Therefore, a combination of structural, waterproofing, or corrosion control methods may be necessary at the same manhole. Some manufacturers advertise products that meet all three requirements. The broad categories of manhole rehabilitation are chemical grouting, cementitious manhole restoration, polymers (epoxy, polyurethane, and urea), cured-in-place liners, chimney seals, barrel joint seals, and precast inserts. See *NASSCO Performance Specification Guidelines for Renovation of Manhole Structures*, for more detailed information (http://www.NASSCO.org).

2.9.3 Channels

Smooth, continuous manhole channels efficiently move wastewater from one or more inlet lines to the outlet lines. Properly designed and constructed side or angle inlets prevent deposition of solids at the transition point.

Wherever a change in pipe diameter occurs, the pipe crowns or water surfaces are located at the same elevation to reduce flow resistance. Where there is a change in direction without a change in diameter at a manhole, a drop in elevation is allowed through the manhole. The following drops are recommended: a 0- to 45-deg angle change over a distance of 3 cm (0.1 ft) and a 46- to 90-deg angle change over a distance of 6 cm (0.2 ft).

Inlet pipes that are too high to enter the manhole at the bottom (more than 31 cm [1 ft]) can enter through a drop connection. However, every effort should be made to avoid sudden drops in flow. Sudden changes in flow direction will cause gases such as hydrogen sulfide to be released out of solution.

Constructing shelf areas so that they drain to the channel at a slope of at least 1:12 reduces deposition. Because of careless construction, a channel of good hydraulic properties, though important, is not frequently created. The channel should be, as far as possible, a smooth continuation of the pipe. To this end, the pipe is sometimes laid through the manhole with the top half removed to provide the channel (ASCE and WEF, 2007). The channel walls should be formed or shaped to three-fourths the full height of the crown along interceptor sewers and other manholes with high velocities. This practice will help reduce deposition of solids on the shelf and facilitate smooth flow hydraulics through the manhole. Making the channel wall the full height for most 0.2-m (8-in.) diam sewers increases the difficulty of sanitary sewer investigations and maintenance tasks such as:

- Physical inspection of pipes by lamping,
- Insertion of television inspection cameras,
- Deployment of cleaning equipment,
- Insertion of sewer plugs, and
- Installment of some flow measurement equipment.

2.9.4 Covers

Manhole frames and covers are constructed of heavy cast iron or ductile iron. They need to be approximately 60 cm (24 in.) in diameter to allow easy access. Machining the horizontal bearing surfaces between the cover and frame reduces rocking and provides a tighter fit, especially if a rubber gasket is grooved into the frame or lid.

In flood plains, flood-prone areas, or areas subjected to ponding or sheet flow runoff, manhole covers should be watertight (with or without bolts). These covers and rubber gaskets are designed to eliminate inflow. Lifting holes or bars in the lid are designed to be non-penetrating to reduce surface water inflow. Removal methods that do not require lifting holes are preferred; however, if lifting holes are necessary there should be no more than two. Alternatives to replacing an existing hole-penetrating cover are either hole plugs or frame and cover inflow inserts (dish). The inserts stop or impede the inflow from entering the system through the cover penetrations. The inserts are made of plastic, polyethylene, or metal materials.

Where manholes are located in poor soils, shifting sands, or soft asphalt pavement on a hot day, it is common to install a concrete ring that extends approximately 0.3 m (1 ft) around the top of the manhole frame and cover. This provides maintenance crews with firm footing for a confined space tripod and for lifting bars. A secondary benefit of the concrete ring is that it makes the manhole easier to locate both in drive-by searches and in aerial photographs.

Adjustment rings between the cone or slab and the rim casting may be appropriate in manholes located in streets and alleys. These rings facilitate adjustment for regrading projects. It is important to take care to waterproof and seal the joints between the manhole frame, adjustment rings, and cone because this is often the area where inflow enters the manhole.

In cold climates or where traffic loads are heavy, adjustment rings may fail and allow I/I or loss of backfill. Rings of high-quality concrete may be used to help avoid this. Flexible frame-to-cone seals that permit movement from frost action or traffic effects are a viable alternative.

2.9.5 Steps

Corroded or poorly grouted manhole steps (rungs) present a safety hazard. These steps can break off underfoot, possibly causing a serious or fatal fall. Efforts should be made to remove steps as part of a rehabilitation program. Most agencies are dispensing with steps in new manholes and equipping maintenance vehicles with lightweight aluminum or fiber-glass extension ladders or portable hoisting frames and harnesses.

2.9.6 Drop Manholes

Drop manholes are manholes where the vertical distance between the influent pipe and the effluent pipe is greater than 0.3 m (1 ft). Drop manholes that have a drop greater than approximately 0.6 m (2 ft) should have two influent pipes. The top pipe enters the manhole on the same grade as the upstream sewer pipe. The bottom pipe has the same vertical alignment as the upper pipe but comes into the manhole at the invert of the manhole. There are both outside (most prevalent type) and inside (less favored) drop manholes used in sewer systems. The design again depends on the jurisdiction and their experience.

Drop manholes typically are considered when it is not cost effective to make an influent line match the effluent line. An example of this is when the construction cost savings of installing the sewer pipe at the extra depth exceeds the cost of the extra piping and workmanship of the drop connection at the manhole. However, because drop manholes can create other costs associated with materials corrosion and maintenance problems, other approaches are often used. For more information, see *Gravity Sanitary Sewer Design and Construction* (ASCE and WEF, 2007).

2.9.7 Connection between Manhole and Sanitary Sewer

Design should anticipate the possibility of differential settling of the manhole and the sanitary sewer pipe. Design and construction precautions are warranted where the manhole and sewer connect. Shear cracks or stresses in a sewer pipe can occur at or immediately outside the manhole. The use of flexible pipe-seal connections integral to the manhole wall is a method typically specified in design to avoid buildup of stress forces in the pipe. If flexible pipe-seal connections are not applicable or used, a flexible pipe joint at or within approximately 1 m (3.3 ft) of the outside manhole wall decreases the likelihood of pipe damage.

Designers should allow for the change in trench width as the pipe enters and leaves the manhole. Also, the contractor should properly place, compact, and support the pipe with an appropriately graded granular bedding and backfill material. Otherwise, the sewer pipe may develop shear stresses and fail.

Corrosion problems may be anticipated where a force main discharges to a manhole as discussed previously. Hydrogen sulfide (generated principally by the bacterial reduction of sulfate in wastewater) is released, oxidized, and condensed on exposed surfaces as sulfuric acid. This process can cause extreme damage to cement, carbonaceous aggregate, asbestos-cement, unlined ductile iron, and steel. Limiting turbulence at these juncture points will reduce the release of hydrogen sulfide gas.

2.10 Manhole and Pipeline Testing

2.10.1 New Construction

Emphasis on eliminating I/I from the collection system has spurred the enforcement of performance tests that allow only relatively minor amounts of infiltration or exfiltration. Hydrostatic tests typically are used for pressure pipelines, manholes, and, to a lesser extent, small-diameter gravity sewers. Pressures for pipelines are established based on design requirements of the pipeline system and should represent the greatest pressure to be generated in the pipe. Pressures for manholes are static pressure only; the manhole is filled with water and the rate of exfiltration is measured.

Air testing typically is used for gravity sewers and manholes. However, manhole pressure tests are negative-pressure or vacuum at approximately 14 to 21 kPa (2 to 3 psi). Pressure increases over time are measured against acceptable rates.

In addition to air or hydraulic tests, it is a good practice to pull a mandrel through the installed pipe or to require CCTV inspection to note any observed leakage or defects. Mandrels are used to determine whether an installed pipe has deflected more than accepted by the designer. Mandrel tests typically are used with plastic pipe installations because they will deflect easily if not properly backfilled. Two basic types of mandrels are available. One is a rigid mandrel constructed with a fixed deflection that, if exceeded, will hedge in the pipe and not advance in the pulled direction. The other is a mandrel with telescoping prongs that are depressed by a deflected pipe and, when removed from the pipe, are measured against the designer's allowable deflection criteria.

Excessive deflection reduces flow capacity and can worsen over time. Television inspection for new construction can reveal manhole and pipe leaks or defects that may exist but that do not cause an air or hydraulic test failure. Also, CCTV can reveal line sags or sediment or debris left in the pipe after construction.

Some communities require each manhole and pipe to be tested or CCTV inspected. Others establish a minimum percent to be tested or televised with provisions for more if specified criteria are not met. Using CCTV to inspect the entire system before acceptance (and before paving above the pipe) not only confirms that the system is installed correctly, but provides documentation for future comparisons.

2.10.2 Rehabilitation

The same tests for new pipelines and manholes apply to rehabilitated pipelines and manholes. However, rehabilitation of manholes should include additional tests typically not required for new construction (except for the previous cases). These tests are a surface density test (hammer test), an adhesion test, and a pH test.

Surface preparation for manhole rehabilitation is a key factor in achieving adhesion of the structural or waterproofing liner to the manhole substrate surface. Surface density tests, pH tests, and petrographic tests (ASTM C856-04) can provide important information about the condition of the substrate material (ASTM, 2004).

Petrographic tests performed by a qualified person can reveal whether the concrete composition and the surface are cleaned properly. The adhesion test is an indicator test that ensures that the complete liner is properly applied.

2.11 Inverted Siphons

An inverted siphon is a section of sewer that, because of a large obstruction such as a river, depressed freeway, subway, or utility line, must be installed below the hydraulic gradient and thus operates under positive pressure. Most inverted siphons (or depressed sewers) contain at least two pipes (barrels) although three pipes are preferred: a smaller pipe for low flows to provide greater velocity for scouring, a midsize pipe for larger flows, and a larger barrel to carry peak or future flows. To eliminate overflow or backup conditions, some inverted siphons include another pipe capable of carrying the entire flow in an emergency.

Inverted siphons typically have maintenance access structures at each end. Gates or stop plates at the inlet structure allow the diversion of all flow through any one barrel for scouring. Flap gates held open by a chain and accessible from the top of the manhole have been used successfully in small siphons. Working space is allotted in the inlet structure to launch a cleaning pig and in the outlet structure for recovery. To allow passage of cleaning equipment, smooth curves of adequate radius (less than 22 deg) are preferred. Inverted siphons have been successfully constructed with various types of pipe, including ductile iron, concrete, and polyethylene.

Odor problems tend to occur at the inlet chamber because air is released by movement induced by wastewater flow and lack of a free-water surface through the siphon. Positive pressure develops in the sewer upstream of a siphon because of the downstream movement of air induced by the wastewater flow. Air tends to exhaust from the manhole at the siphon inlet and can cause odor problems. The problems typically are cured with air lines or air jumpers.

The inlet chamber has a lateral overflow weir and, sometimes, a sump to trap grit and heavy objects and to prevent clogging of the barrels. Screens may be installed if the added expense of periodic cleaning is warranted. Inlet and outlet structures both need tamper-proof manhole covers.

Both the inlet and outlet structures are prime environments for hydrogen sulfide release, which results in an aggressive corrosion attack on concrete and iron materials. Corrosion control is vital for these facilities if a reasonable life expectancy is desired.

2.12 Sanitary Sewers above Grade

Sanitary sewers are sometimes placed over an obstacle where topography, width of crossing, and flood elevations and velocities allow. They are installed or hung on bridges, supported on piers, suspended by cable spans, or fastened to a structural beam. Pressure-type joints, insulation, and jacketing prevent leakage and protect the pipe against freezing, although above-grade sewers are rarely used in cold areas.

2.13 Crossings and Tunnels

2.13.1 Stream Crossings

Sanitary sewers that transect a stream and are installed using traditional excavation techniques often have pipe tops that are several meters beneath waterway inverts. This typically is done by necessity because there is minimum slope available, forcing the pipe to be as shallow as possible. These crossings are subject to erosion and washout because of their proximity to the stream invert. Crossing pipe can be encased in concrete or placed in another pipe to protect against scour, effect, and potential inflow. Pipe joints designed specifically for crossing bodies of water also are available. If the crossing is complex, high risk, or expensive, alternative crossing options such as microtunneling or directional drilling should be evaluated. Regulatory agencies increasingly are requesting tunneling to avoid concerns with open-cut techniques.

Some agencies require that crossing signs be posted on either side of a navigable stream or lake that are visible from land and water. Signs should alert the public to the presence of the submerged pipe so that activities such as dredging, geotechnical explorations, and boat anchoring do not penetrate or shift the pipe's position and make it more vulnerable. A contact phone number and any other special notes should be legible from a reasonable distance as determined by the specific site conditions.

Substantial problems may be encountered where sanitary sewers span a watercourse, particularly when support piers are required. Brush and debris invariably jam against the pipe, sometimes causing a rupture and discharge of untreated wastewater to the stream. Bridged crossings ideally are placed above the maximum flood level. Stabilizing stream banks helps to reduce erosion maintenance. Manhole structures should be protected or physically located to reduce washout by bank erosion.

2.13.2 Other Crossings

There are many other cases where pipelines need to cross major highways and busy intersections or to avoid public concerns. In these cases, tunnels or trenchless methods are being used more often. The method of tunneling depends primarily on the diameter of the crossing and the soil conditions.

When crossing major freeways, agencies will require an encasement pipe to provide double protection to their facilities. This is especially true when using pressurized piping.

2.14 Service Connections and Disconnections

Service connections also are referred to as house connections, building sewers, service lines, or service laterals. They connect the customer's facility (building) to the public sewer system. Service connection requirements for each community vary as to the criteria for design and installation.

Attention to service connections has greatly increased over the past 10 to 15 years because of the recognized difficulties with I/I. Because the length of service connection piping can equal or exceed the length of public property sewers, their contributing effect on wet-weather flows can be significant. This fact recently prompted a change to the long-held cultural restraint of investing public money on private service connections; public money may be more effectively spent to remove defects. Regardless of the approach to correcting service connection defects, designers of sewers should include realistic flow allowances for I/I from service connections.

Optimum sanitary sewer design includes properly located and securely plugged Y and T connections for future service connections. Typically, the public sewer is installed and then the service Y or T connections installed with a section of service pipe plugged at the end before extending to the private property boundary. Close attention needs to be given to designing a tightly plugged pipe; otherwise, the service connection can become a significant source of I/I and sediment. Inexpensive mechanical plugs are alternative. Also, accurate location records help make Y and T connections easily recoverable when the plumber wants to complete the building connection or perform repairs.

Sometimes, service connections are not available on the sewer at the desired location. Service connection installation via the break-in or hammer-tap methods can cause pipe failure and line blockage and makes pipes vulnerable to infiltration and root intrusion. Local regulations typically prohibit break-in service connections and require machine-tapping and saddle-style or other suitable factory-fabricated connections. It is strongly encouraged that construction inspection is performed on service laterals at the time of installation.

Old structures may be razed and the excavated land backfilled during property development. Typically, sanitary sewer service is not properly disconnected, and the construction design will require a new service connection. Several generations of site reuse may result in multiple service connections. Inactive connections are a source of infiltration, insect and rat infestation, and other possible maintenance problems. Therefore, building permit and design regulations often may require that abandoned services be disconnected and permanently plugged at the main, if possible. Again, ensuring a watertight plug and an accurate location typically will restrict the potential problems that could otherwise result.

As stated above, more attention is being paid to service laterals or private property sewers. Many communities are now ensuring that private systems are being rehabilitated as required to reduce the amount of I/I. One approach is requiring inspection and upgrades at the time of property sale or ownership transfer. Another is to have owners pay monthly into an insurance policy that can be used to rehabilitate laterals. Sharing the cost of lateral rehabilitation between the owner and the utility also has been used successfully in some communities.

A resource called the Private Property Virtual Library has been created to address ongoing developments in this area (http://www.wef.org/apps/PPVL_Site/wef/ppvl_main_page.asp).

2.15 Value Engineering

Use of value engineering has typically been set aside for large, multi-faceted projects. The intent is to draw together experts in the areas the project addresses. For instance, a large sewer project with numerous tunnels and horizontal directional drilling (HDD) might have a team composed of a sewer designer, a geotechnical engineer, a tunnel designer, a HDD specialist, a cost estimator, and certified leader. It is recommended to have a leader who has served on teams before and knows how to get to the end of the assignment.

2.16 Specification Writing

Specifying the type of materials that the project needs and their minimum performance criteria is important to project success. It is recommended that designers use the Construction Specifications Institute (CSI) standard format and guidelines when developing a set of technical specifications. The CSI is divided into many divisions that each deal with certain aspects of construction. Each division is further separated into specific sections dealing with particular items such as pipe material or testing. The format is straightforward and provides general information, references, descriptions of products and materials, and installation requirements and testing.

3.0 REFERENCES

American Society of Civil Engineers; Water Environment Federation (2007) *Gravity Sanitary Sewer Design and Construction,* 2nd ed.; ASCE Manuals and Reports on Engineering Practice No. 60; WEF Manual of Practice No. FD-5; American Society of Civil Engineers: Reston, Virginia.

ASTM (2004) *Standard Practice for Petrographic Examination of Hardened Concrete;* ASTM Standard C856-04; ASTM International: West Conshohocken, Pennsylvania.

National Association of Sewer Service Companies; *NASSCO Performance Specification Guidelines for Renovation of Manhole Structures.* http://www.NASSCO.org (accessed March 2008).

Private Property Virtual Library. http://www.wef.org/apps/PPVL_Site/wef/ppvl_main_page.asp (accessed November 2008).

Water Environment Federation (1993) *Design of Wastewater and Stormwater Pumping Stations;* Manual of Practice No. FD-4; Water Environment Federation: Alexandria, Virginia.

Water Environment Federation (2007) *Wastewater System Capacity Sizing Using a Risk Management Approach;* Technical Practice Update. http://www.wef.org/ScienceTechnologyResources/TPUs/; May.

Water Environment Federation (2008) *Alternative Sewer Systems,* 2nd ed.; Manual of Practice No. FD-12; Water Environment Federation: Alexandria, Va.

Water Environment Federation; American Society of Civil Engineers (2008) *Existing Sewer Evaluation and Rehabilitation,* 3rd ed.; Manual of Practice No. FD-6; ASCE Manuals and Reports on Engineering Practice No. 62; McGraw-Hill: New York.

Chapter 6

Construction Contracting

1.0 CONSTRUCTION PROJECTS AND PROJECT MANAGEMENT 138	5.2 Term Contracts 145
	6.0 CONSTRUCTION CONTRACT DOCUMENTS 146
2.0 LAWS AND REGULATIONS 139	6.1 Constructability Review 146
2.1 Contracting Requirements 139	6.2 Operations and Maintenance Review 146
2.2 Competitive Bidding 140	6.3 Value Engineering 146
2.3 Environmental Requirements 141	6.4 Allocation of Risk 147
2.4 Interagency Permit Requirements 141	*6.4.1 Unforeseen Conditions 147*
	6.4.2 Dispute Resolution 148
3.0 CAPITAL PROJECT DELIVERY METHODS 142	*6.4.3 Conflict Resolution 148*
3.1 Design-Bid-Build 142	*6.4.4 Escrow Documents 148*
3.2 Multiple Prime Contractors 142	7.0 FINANCIAL CONSIDERATIONS 148
3.3 Design-Build 143	7.1 Payment Methods for Construction Contracts 148
3.4 At-Risk Construction Management 144	*7.1.1 Lump-Sum Basis 149*
4.0 AGENCY CONSTRUCTION MANAGEMENT 145	*7.1.2 Unit-Price Basis 149*
	7.1.3 Cost-Plus Basis 149
5.0 MAINTENANCE PROJECT DELIVERY METHODS 145	7.2 Funding Mechanisms 150
5.1 Force Account Construction 145	*7.2.1 Municipal Bonds (Local Financing) 150*
	7.2.2 Federal and State Financing (Grants and Loans) 150

(continued)

7.3	Cash Flow/Cost Control	151		Closed-Circuit Television Inspection	153
7.4	Construction Bonds	151	8.4	Warranty/Guarantee Administration	154
7.5	Construction Insurance	151			
8.0	QUALITY ASSURANCE/ QUALITY CONTROL	152	9.0	SAFETY CONTROL METHODS AND PROCEDURES	154
8.1	Construction Inspection/ Administration	153	10.0	UTILITY DAMAGE CONTROL (ONE-CALL SYSTEMS)	154
	8.1.1 Field Directives	153	11.0	ALLIANCES AND PARTNERING	155
	8.1.2 Engineering Bulletins	153			
8.2	Schedule	153	12.0	REFERENCES	155
8.3	Post-Construction		13.0	SUGGESTED READINGS	157

1.0 CONSTRUCTION PROJECTS AND PROJECT MANAGEMENT

This chapter provides an overview of public works construction contracting for wastewater collection systems and provides sources of more detailed information. Significant information and guidance also are available from the following organizations: Construction Specifications Institute (CSI); Engineers Joint Contract Documents Committee (EJCDC); Construction Industry Institute (CII); American Public Works Associations (APWA); American Institute of Architects; Associated General Contractors of America; Associated Builders and Contractors; National Utility Contractors Association; and American Society of Civil Engineers.

Wastewater collection systems construction projects involve complex processes. Even small projects involve different skills, materials, equipment, and operations. The construction process is subject to variable and unpredictable factors. Many aspects of construction sites contribute to the complexity of construction projects: subsoil conditions, surface topography, weather conditions, transportation systems, material and equipment supply, availability of skilled labor, local subcontractors, and many others. Therefore, to some degree, every construction project is unique.

A construction project typically involves many parties including the owner, design engineer, prime contractor, subcontractors, manufacturers, material and equipment suppliers, inspection and testing laboratories, and surety and insurance companies. The owner of the structure finances the project. The structure is designed in accordance with applicable codes and standards and follows state and administrative directives regarding advertising, bidding, contracting, and administration of the construction process. For competitive bidding, the owner defines the scope of the project, describes the materials and tolerances, develops complete and detailed descriptions of the proposed work, and defines the conditions by which the work will be performed. The design engineer describes these elements in drawings and specifications referred to as construction contract documents.

Construction drawings illustrate the extent and arrangement of the components of the structure. The specifications describe the materials and workmanship required. Construction contract documents are either prepared in-house or subcontracted to an independent design engineer, depending on the complexity of the project and the owner's design capabilities.

Project management is the overall control of design and construction to achieve an optimum balance between quality, schedule, and cost. Project management, however, does not exert a constant, uniform influence over these attributes throughout all phases of project delivery. Project management influence is greatest during definition and preliminary planning phases and tends to decrease rapidly as the process continues toward design and construction. Significant decisions concerning overall project size, scope, complexity, location, and performance characteristics are made during the early stages of the project. After the early stages of the project, firm commitments of project resources have to be made. As the owner commits resources to the project—such as acquiring facility sites, engaging consulting engineers for detailed design, and hiring contractors to perform construction—it becomes more difficult and expensive to make changes to the project plan. As a result, proper planning and decision-making are extremely important during project definition and preliminary planning.

Software is available to provide the tools and resources for project management. These project control systems include scheduling, estimating, and providing information management to control the budget, schedule, and quality of the project.

2.0 LAWS AND REGULATIONS

2.1 Contracting Requirements

There are several items to consider before a project is bid. Special laws such as building codes, environmental regulations, building permits, safety and health regulations, zoning regulations, prevailing wage requirements, and licensing for contractors may be applicable. These are intended to protect public health and safety and prevent fraudulent acts by owners and contractors. Contractors should be informed of these requirements in the bidding process.

Many states have a requirement that all public works projects be designed by a licensed professional. Some set a dollar limit after which a licensed professional must prepare the design. For example, the New York State Education Department states, "No county, city, town, or village or other political subdivision...shall engage in the construction or maintenance of any public work" that is greater than $5,000 unless reports, plans, and specifications are prepared by a licensed professional engineer (NYS, 2008). It is important to note that public works "maintenance" in New York requires involvement of a professional engineer. Kentucky, on the other hand, does not set a dollar limit for public works projects and specifically excludes "maintenance or repair of the facility" (KLRC, 2007). Texas sets a dollar value at greater than $8,000, if involving mechanical or electrical engineering, or $20,000 if mechanical or electrical engineering is not involved (TBPE, 2008). Texas appears to exclude public works maintenance.

New York has a unique provision that requires separate contracts for general construction; plumbing; heating, ventilation, and air conditioning; and electrical for any public works building with a value greater than $500,000 to $3 million (varies by county) (NYS, 2008). A project labor agreement (PLA) can waive the provision for separate contracts. There are other specific criteria that must be met when establishing a PLA, such as an apprenticeship program. Public works managers need to identify and understand the various contracting requirements that are applicable to their state and particular jurisdiction and their implications on the project.

2.2 Competitive Bidding

Laws or regulations typically mandate the competitive bidding of utility projects, ensuring value to ratepayers and fairness in the expenditure of public money. Public administration of bids typically is required for publicly funded work. Project bids, which are based on construction contract documents, are designed to receive cost-competitive bids and to provide the owner with a constructed project of specified quality at the lowest possible price.

Care should be taken when preparing public works bid documents to avoid language that is ambiguous or allows apparent "owner discretion." For example, the bid documents for a public works project in Minnesota included a statement that the owner had a right to "waive informalities or irregularities in a Bid received and to accept the Bid which, in the Owner's judgment, is in the Owner's own best interest." After the bids for the project were received and opened, the second lowest bidder indicated that it made a human error on its bid, and asked the owner to change the value of one of the bid alternates from a plus (+) sign to a negative (-) sign. The owner changed the sign under the pretense that this was a bid irregularity that it could waive. The second bidder then became the low bidder and was awarded the contract. The original low bidder filed a bid protest. Although the original low bidder was not awarded the contract, it did prevail in Minnesota Supreme Court, which stated that (Knoll, 1997): "A fundamental purpose of competitive bidding is to deprive or limit the discretion of contract-making officials in the areas which are susceptible to such abuses as fraud, favoritism, improvidence, and extravagance. Any competitive bidding procedure which defeats this fundamental purpose, even though it be set forth in the initial proposal to all bidders, invalidates the construction contract although subsequent events establish...that no actual fraud was present." Owners should be certain that their general provisions are adequately reviewed by appropriate legal professionals specializing in public works law. Owners should then be certain to scrutinize contract documents that it will use for bidding to ensure that ambiguous or contradictory language is not included in special conditions or technical specifications.

2.3 Environmental Requirements

Federal, state, and local governments have many laws and policies to protect environmental quality. Most have a direct effect on a public works project, such as permitting requirements for work near waters of the United States, including wetlands. Construction projects that disturb as little as 0.4 ha (1 ac) of land may require mitigation of site drainage and runoff under the U.S. Environmental Protection Agency's National Pollutant Discharge Elimination System Construction General Permit (U.S. EPA, 2008). Others have indirect effects, such as pending regulations to reduce air emissions from nonroad diesel equipment (e.g., construction equipment). These environmental requirements can affect a public works project at all phases, including lead time to acquire permits, stop-work orders by agencies having jurisdiction for noncompliance with permit requirements, and postconstruction monitoring and reporting such as for invasive species of plants in surface restoration of wetlands.

2.4 Interagency Permit Requirements

Interagency permits include any necessary authorization to work in a public right-of-way, either to open cut a road to install a casing pipe by a horizontal bore or for road patching, plumbing, or building (e.g., for pumping stations). Sewer agencies may have blanket permits with other agencies (e.g., a state department of transportation) to allow for routine maintenance of facilities (e.g., cleaning catch basins). But these blanket permits may not apply to capital improvements, particularly when the sewer agency hires a contractor to perform the work. In some cases, the permitting agency may waive a fee if the permit is obtained on behalf of the sewer agency. Review by a historic preservation board or a parks commission also may apply. As with environmental permits, these interagency requirements can affect project schedule. Any necessary mitigation, such as improvements to a park or street, also can affect the project's cost.

3.0 CAPITAL PROJECT DELIVERY METHODS

3.1 Design–Bid–Build

The traditional project delivery method for public works projects is design–bid–build. As the name implies, it is a sequential process. First, the owner prepares a set of contract documents, either with inhouse resources or through a contracted design professional. Then the project is advertised, and bidders prepare estimates and submit bids based on the contract documents. The project is awarded and construction of the project begins. The owner oversees construction either with inhouse or contracted resources.

This method is time-tested and is known to virtually all owners, engineers, and contractors. The product is well defined because all aspects are identified and designed prior to bidding. The design–bid–build process is time-consuming because of the sequential steps of the work. This method can limit the designers understanding of construction labor, scheduling, material availability, and constructability. And it can create an adversarial relationship between the engineer, contractor, and owner. The nature of the process may cause the contractor to look for least-cost approaches to complete the project and meet or exceed its expected profit projections. This, in turn, creates the need for increased oversight and quality control by the owner. Lack of knowledgeable contractor contribution to the design may limit the effectiveness and constructability of the project.

3.2 Multiple Prime Contractors

In this method, the owner has separate contracts with several providers, such as general construction, mechanical, plumbing, and electrical. The owner, with its own resources or through a contracted professional, manages the overall project schedule and budget during the construction phase. Pennsylvania, North Dakota, Illinois, and New York require this for all public works building projects. It allows for "fast-tracking" as each piece is bid separately, allowing for flexibility in awarding contracts for the first portions of the project. This requires that the design for each piece is completed. Fast-tracking can be a highly desirable for procurement in cases where time is a critical element.

There are, however, several disadvantages to this method. The final cost of the project will not be known until the final prime contract is bid and awarded. An absence of overall authority once construction is underway can lead to problems primarily from lack of coordination and contractor delay issues. An owner may try to give the general construction contractor contractual responsibility to coordinate the work among trades, including project schedule. The general prime contractor, however, lacks contractual authority to dictate the schedule of another contractor. Courts have upheld that public owners are responsible for the management of multiple prime contracts (GBCNYS, 1972).

3.3 Design–Build

Design–build replaces the traditional method of awarding separate contracts for design and construction. Design–build is a method of project delivery in which a single entity provides all of the services necessary to both design and construct the project. This single point of responsibility fundamentally distinguishes the design–build format from other forms of project delivery. A single source of responsibility reduces the owner's administration and management burden, protects the owner from liability, and reduces disputes concerning responsibility. Planning and procurement for successful design–build delivery differs in several ways from traditional project delivery. Strong owner involvement is required at the planning stage to determine the performance characteristics of the facility with sufficient clarity to establish fair competition. The procurement process also must allow designers sufficient flexibility to consider economical technical alternatives and require a guarantee for facility performance. It also must allow a basis for fair comparison between alternatives and provide a remedy when performance criteria are not met. Because teamwork is an important part of the process, owners must take great care to select qualified, functional design–build teams. The selection process described for construction management should also be used for the design–build format.

The design–build project delivery system has grown in popularity. It is used to some degree in most states, except in Alabama, Iowa, Michigan, Rhode Island, and Wyoming where design–build is "not specifically authorized for public agencies" (AIA, 2006).

3.4 At-Risk Construction Management

In at-risk construction management, the construction manager is expected to deliver a project to the owner within a guaranteed maximum price (GMP). The construction manager acts as advisor to the owner in planning and design phases and provides scheduling, budgeting, and constructability advice. During the construction phase, the construction manager is the equivalent of a general contractor. Because the construction manager is contractually responsible for a GMP, the nature of the relationship between the owner and construction manager is changed. The construction manager must protect his or her own interests in addition to those of the owner.

At-risk construction management is comparable to the traditional design–bid–build method because the construction manager performs as the general contractor during construction. Specifically, the construction manager holds the risk of subcontracting the various disciplines of the construction work to various trades and guarantees completion of the project for a predetermined, negotiated price following completion of the design. This delivery method is comparable to design–build because construction can begin before completion of the design. The construction manager can solicit bids and subcontract various portions of the work while design of unrelated portions is still being completed.

A disadvantage of this delivery method is the contractual relationship among designer, construction manager, and owner as soon as construction starts. Once construction begins, the construction manager ceases to be an advisory to the owner and assumes the contractual role of general contractor. Quality of construction, issues associated with design, and schedule and budget can become points of contention. At this point, at-risk construction management becomes analogous to the traditional design–bid–build system, and adversarial relationships may result.

A significant benefit of this method is the opportunity to incorporate an experienced contractor's perspective into the decisions associated with planning and design of the project. The project also can fast-track early components before design is complete, saving time and associated costs.

4.0 AGENCY CONSTRUCTION MANAGEMENT

Agency construction management is when the construction manager serves as an owner's primary advisor for a project or program. The agency construction manager is accountable to the owner for managing all or part of the planning, design, construction, and postconstruction phases. The construction manager protects the interests of the owner in its dealings with design and construction professionals and with other entities both private and public. The agency construction manager ensures optimum use of available funds, controls the scope of work, and is responsible for project scheduling. By coordinating the most advantageous use of the design and construction firm's skills and talents, the manager strives to avoid delays, changes, and disputes and to enhance quality. This method allows for optimum flexibility in contracting and procurement.

5.0 MAINTENANCE PROJECT DELIVERY METHODS

5.1 Force Account Construction

Construction projects involving repairs, small jobs, and emergency cases often are administered through a method called force account construction. Under this system, the owner acts as the prime contractor responsible for coordinating the project and either subcontracts work to specialty contractors or elects to do the construction with existing work crews.

5.2 Term Contracts

A term contract is a contract in which a source or sources of supply or service are established for a specific period of time at a predetermined unit price. For example, a sewer repair term construction contract could consist of per-unit prices for the in-kind replacement of various sizes of sewer. It might cover, for example, sizes of sewer ranging from 150 to 1050 mm (6 to 42 in.) at various ranges of depths such as 0 to 3 m (0 to 10 ft) and 3 to 5 m (10 to 16 ft), with a variety of surface conditions such as unpaved, light asphalt, heavy asphalt, or concrete pavement. As the utility project manager identifies individual defects in various pipe segments throughout the collection system, such as by closed-circuit television (CCTV) inspection, he or she can accumulate a list of required repairs. The project manager can then issue work orders to the contractor who holds the term construction contract to replace the sections of pipe with the defects using the established unit prices.

6.0 CONSTRUCTION CONTRACT DOCUMENTS

6.1 Constructability Review

As outlined in Section 3.0 Capital Project Delivery Methods, an advantage of the construction management and design–build approaches is the innate contribution of a knowledgeable construction contactor. When using the traditional design–bid–build, force-account-construction, or term-construction approaches, the owner should have the design reviewed by an independent individual or firm with good working knowledge of the construction methods applicable to the project. The owner should be aware, however, that a contractor who performs a constructability review may become ineligible to bid on the project. The reviewing contractor may be considered to have an unfair advantage over other contractors.

6.2 Operations and Maintenance Review

The operations and maintenance (O & M) staff should have an opportunity to review and comment on the project at various stages. This real-world input by O & M staff provides the same benefits to the long-term viability of the project as contractor input has to constructability. The O & M staff bears the burden of design and construction errors long after the engineer and contractor have completed the project. O & M staff can assist with the development of life-cycle costs that can be used to evaluate capital improvement alternatives.

6.3 Value Engineering

Value engineering typically involves engaging a separate engineering group to examine a project plan (or conceptual design) to question whether it will in fact deliver the value that the owner wants in a cost-effective manner and to consider whether alternatives will prove more reliable and cost effective. The process forces the owner to question and determine real needs. It also assists in identifying the main advantages and disadvantages of alternative engineering options. Properly conducted value engineering studies should end with general agreement on the best engineering design for a particular owner, based on whatever the owner values most.

Value engineering can be introduced during the construction phase of a project. Because it is more difficult and expensive to change plans after significant resources have been committed to an existing plan, value engineering has been found to deliver the most value for the least cost during the earliest stages of a project.

Many construction contracts with public agencies include value engineering incentive clauses. Value engineering uses the contractor's special knowledge to reduce the cost of a project without sacrificing quality or reliability. A clause of this type provides the contractor with an opportunity to suggest changes in the plans or specifications (such as changing, for example, from conventional trenching to trenchless technology) and share the resulting savings.

6.4 Allocation of Risk

Risk is the probability of the occurrence of a defined hazard (i.e., a particular event which has the potential if it occurs of an adverse effect) and the magnitude of the consequences. Risk assessment is the estimation and evaluation of the risk; the magnitude of the consequences together with the probability of the consequences. Risk identification is an awareness of those risks that could adversely affect the outcome of the venture or project. Risk management is the identification, measurement, and economic control of risks.

There are several factors that present risk to the owner, contractor, and engineer including errors and omissions in design; environmental, health, and safety; weather; quality of work; subcontractor performance; and inadequate project definition or organization. The utility project manager should discuss with its design professional or construction manager, or both, his or her tolerance for risk. This will help drive decisions on the level of effort for project components such as sub-surface investigations, payment methods, and umbrella insurance policies.

6.4.1 Unforeseen Conditions

Unforeseen conditions are one area in which an owner can affect risk and cost. Sewer projects are risky and the potential for unforeseen conditions along any portion of the length of a project can have profound effects. If an owner is risk adverse and states in its contract documents that (1) the owner accepts no responsibility for subsurface conditions; or (2) the contractor assumes all liability for all conditions found at the site whether or not they could have been reasonably anticipated, the owner is not only putting a high degree of risk on the contractor but is placing a financial burden on itself. If the contractor perceives an undue risk, this will be reflected in its bid prices. The owner should take the necessary steps to give the contractor as much information as possible, via a thorough subsurface program, or should be willing to share the risk of truly unforeseen conditions. The allocation and management of risk will be reflected in the cost of construction.

6.4.2 Dispute Resolution

The Construction Industry Institute states (CIIDPRRT, 1995), "Disputes arise when the field project managers are unable to resolve a disagreement and others (e.g., project sponsors, advisors, neutral parties) become involved in an attempt to facilitate agreement." Disputes are identified as occurring on the jobsite and are a function of uncertainties surrounding the project, problems with process (e.g., incomplete scope definition), and issues involving people (e.g., poor interpersonal skills). Involvement of "others" who are not involved in day-to-day activities, yet share the goal of successfully completing the work, can be an effective method of resolving a dispute. Neutral parties or advisors can be individuals who are mutually selected by the owner and contractor before construction begins. This "disputes resolution board" is charged with rendering a decision based on the merits of each parties' interpretation of the dispute and not based on an allegiance to a particular party. The dispute resolution board may meet several times over the projects duration to address routine disputes.

6.4.3 Conflict Resolution

If a dispute cannot be resolved onsite, the conflict is addressed away from the project using techniques such as fact-based mediation, nonbinding or binding arbitration, a minitrial, or private judging. Whereas a dispute resolution board is part of construction phase, conflict resolution is typically an issue-specific technique.

6.4.4 Escrow Documents

On large, complex, or risky projects, such as a sewer tunnel, the contractor's bid documents can be held in escrow by an independent third party. If a conflict arises where the contractor is claiming extra costs, its bid documents can be reviewed to ensure that there was a legitimate omission or anticipation of certain conditions when the bid was being prepared. The placement of bid documents in escrow needs to be specifically identified in the contract documents.

7.0 FINANCIAL CONSIDERATIONS

7.1 Payment Methods for Construction Contracts

Although there are many different types of construction contracts, they may all be grouped within the following significant divisions:

- Contracts for which the contractor is selected based on lowest responsive and responsible competitive bidder. This procedure is typical for public works contracts. Typically, these contracts will be lump-sum and unit-price contracts.

- Contracts resulting from direct owner–contractor negotiations. These can be determined on any mutually agreeable basis, such as lump-sum, unit-price, or cost-plus fee.

7.1.1 Lump-Sum Basis

Lump-sum contracts stipulate a fixed price for project construction. The lump sum reflects the total cost of the contract project. The owner benefits most from this type of contract because the total cost is known in advance. However, lump-sum contracts limit the scope of work because all quantities and complete plans and specifications must be accurately known when the bid is made. The contractor assumes full risk because the job must be completed even though the total cost of the work may be greater than the contract price. Lump-sum contracts are often used in wastewater collection system construction for pumping stations and other well-defined structures.

7.1.2 Unit-Price Basis

Unit-price contracts offer advantages for projects involving quantities of work that cannot be accurately forecast. In this contract, the estimated quantities are compiled by the design engineer, and the contractor bids the unit costs for carrying out the proposed work according to the contract documents. The contractor is obligated to complete the work at the quoted unit price per listed item. Because payments to contractors are based on units of work actually performed and measured in the field, the exact cost of the construction is not known until the project is complete. As a result, unit-price contracts require complete plans and specifications. The contractor assumes less risk with unit-price contracts than with lump-sum contracts. Unit-price contracts often are used in wastewater collection system construction for pipelines and appurtenances.

7.1.3 Cost-Plus Basis

Cost-plus contracts are used when the owner wants a particular contractor to do the work, when the scope and construction techniques are not known, or when expedited project completion is necessary. In a cost-plus contract, the owner pays all the costs plus an agreed management fee; thus, the contractor takes the least risk. The owner and contractor typically negotiate the contract scope. Then, based on preliminary drawings and specifications, the contractor estimates the price. This type of contract often is used in force account work such as emergency repairs for wastewater collections systems.

Examples of items typically included in the management fee are contractor's head office costs, rents, taxes, insurance, interest on financing for the project, head office supervision and control costs, and contractor's profit. Under this type of contract, the contractor has the least incentive to keep costs down, though variations of this contract have been developed to give the owner some cost control and provide incentives to the contractor to reduce costs. These variations include cost plus a percentage of cost, cost plus a fixed fee, cost plus a sliding fee, and guaranteed maximum cost plus cost savings sharing incentives.

7.2 Funding Mechanisms

7.2.1 Municipal Bonds (Local Financing)

Municipal bonds are used to pay for public projects such as the construction or improvement of water and sewer systems and other public works. There are different types of municipal bonds, including general obligation bonds, limited and special tax bonds, industrial revenue bonds, revenue bonds, housing bonds, moral obligation bonds, double barreled bonds, tax anticipation notes, bond anticipation notes, and revenue anticipation notes. The primary difference between these bonds is in how and when the issuer will repay the bonds and how it makes the interest payments. In the case of general obligation notes, the bond is simply backed by the "full faith and credit" of the issuer. That means the local or state government that has issued the bond can use just about any means available to guarantee payments, including raising taxes.

Other bonds are issued with specific provisions to raise taxes or create a new tax. These are known as limited or special tax bonds. If the project being undertaken will generate revenue from tolls, wastewater fees, water bills, or other services, they are known as revenue or industrial revenue bonds.

7.2.2 Federal and State Financing (Grants and Loans)

There are a variety of small-scale federal (e.g., U.S. Department of Agriculture, Rural Utilities Service) and state (e.g., North Carolina Clean Water Management Trust Fund) grant programs. The Water Pollution Control Revolving Fund, also known as the Clean Water State Revolving Fund (SRF), is the largest source of loans for both point source (wastewater and stormwater) and nonpoint source water pollution control projects (U.S. EPA, 1987). The SRF has provided more than $4.5 billion annually in recent years (U.S. EPA, *Wetlands (Section 404) Compliance Monitoring*, http://www.epa.gov/compliance/monitoring/programs/cwa/wetlands.html). More information on project funding and financing options is provided in Chapter 8.

7.3 Cash Flow/Cost Control

Cash flow is important for the contractor and engineer, who rely on revenue to pay staff, buy equipment, and earn a profit. But cash flow projections also are important for the owner who must borrow substantial amounts of money to fund a project. If the owner is undertaking a long-term program or multiyear project, it wants to limit borrowing to only that amount that is necessary to fund a particular portion of the project. In some cases, the owner will ask the contractor to provide cash-flow projections based on anticipated construction production rates and unit prices.

Cost control is important to all parties. Again, the contractor and engineer need to control costs to ensure a profitable job. The owner needs to track costs to ensure that it has sufficient funds to complete the project.

7.4 Construction Bonds

A bond is a legal document that binds a third party to the contract to provide assurance that the contractor will perform the agreed service. Under the bond, a party known as the surety assumes liability for the failure and debt of another. Bonds typically required in construction include bid bonds and contract bonds (performance bonds and payment bonds). Bid bonds assure the owner that the bidder, if selected for award, will sign the contract. Contract bonds guarantee that the contractor will complete the work according to contract documents. If the contractor fails to fulfill the contractual obligations, the surety completes the contract and pays all the costs up to the face value of the bond. Standard bond forms are available from many different organizations, such as CSI and EJCDC.

Construction contracts provide that the prime contractor warrants the work against faulty workmanship and materials for some period after project completion. One year is a typical warranty period, although periods of up to 5 years or more are typical on certain projects such as utility construction. During the warranty period, the contractor remedies construction defects at no cost to the owner. To ensure compliance with the warranty, a type of surety bond called a maintenance bond may be required. A maintenance bond requires the surety, if necessary, to correct defined defects in the contracted work that occur within the warranty period.

7.5 Construction Insurance

Utilities and states typically have specific insurance requirements in their regulations to protect the public; these are incorporated to the construction contract.

For example, construction contracts typically require the contractor to provide coverage for workers compensation insurance, the contractor's public liability and property damage insurance, and the contractor's contingent liability insurance. Some contracts require the contractor to provide property insurance to protect the project and liability insurance to protect the owner. Many contracts require the contractor to protect the owner and design engineer by accepting any liability that may be incurred because of operations performed under the contract. Insurance certificates must often be submitted as proof of insurance.

8.0 QUALITY ASSURANCE/QUALITY CONTROL

Quality assurance (QA) is the sum total of management actions, strategies, and reporting functions required to assure the project owner that a constructed facility satisfies contract requirements and appropriate technical standards. Quality control (QC) is the inspection and testing function required to determine objectively the standard to which work is being performed and facilitates the reporting of results to management. Whereas QA is a management tool, QC is a production tool. Many owners will hire an engineer to develop and manage a QA/QC program. Some agencies have their own QA/QC programs and hire inspection and testing agencies to carry them out. Others require the contractor to develop a QC program, and then assign a person or a third party to provide QA oversight. Elements of the QC program include project meetings, review and approval of submittals, field inspections, materials sampling and testing, and review and approval of record drawings. Systematic documentation and filing are the most important parts of a QA/QC program.

As construction proceeds, testing of individual components is performed, according to manufacturer's recommendations and recognized industry practices, to ensure that no damage will occur during large-scale testing. After construction is complete, total system testing, startup, and commissioning begins. System testing proceeds as required by the manufacturer's recommendations, engineering specifications, and recognized industry practices. Results of these tests are documented carefully and systematically. Adjustments and corrections that may require engineering approval are performed as needed and carefully documented.

After all systems have been tested independently, the complete facility is ready for startup. Startup is performed in accordance with an O & M manual, which incorporates all relevant data from equipment and instrumentation manufacturers, and information provided by the engineer describing how the entire facility is designed to be operated and maintained. After startup, the facility is tested and calibrated for optimum performance and efficiency. In some cases, the engineer may order modifications to the facility to satisfy performance requirements or improve efficiency.

Following commissioning, a final inspection is made, and job files, shop drawings, and O & M manuals (updated to include results of all facility calibration tests and any changes made during commissioning) are turned over to the owner.

8.1 Construction Inspection/Administration

8.1.1 Field Directives

The owner's resident engineer may issue field directives/field bulletins to document extra work within the existing scope of the project that is necessary for satisfactory completion. An example may be the addition of select granular material under the bedding of a pipe to replace unsuitable or soft subsoil. Payment for this work typically is based on existing unit prices or on a time-and-material basis. Field directives or bulletins can be used as the basis for contract modifications and change orders.

8.1.2 Engineering Bulletins

An engineering bulletin typically is issued by the design professional to answer a request for information submitted by the contractor or construction manager. It can also be used to identify a change in the scope of the project and to request a quote from the contractor to complete the new work. Engineering bulletins can be used as the basis for contract modifications and change orders.

8.2 Schedule

A project schedule should be maintained to ensure that the project is completed on time. Properly developed schedules will identify time-dependant tasks that have specific start dates and durations that must be followed to meet the planned completion date of the entire project. Schedules are important regardless of the project delivery method. Schedules can be maintained by the owner, engineer, contractor, or construction manager.

8.3 Post-Construction Closed-Circuit Television Inspection

Internal CCTV inspections should be performed upon completion of installation or repair of sewer systems. A log sheet should be completed to complement the CCTV video tape or video disc. The CCTV inspection is a quality-control measure to ensure that the facility was installed or repaired as specified. It also provides a historical record of the condition of the pipe for use at the end of the guarantee period and during future inspections.

8.4 Warranty/Guarantee Administration

A warranty is a guarantee that a facility will function according to stipulated, agreed performance criteria for a stated period of time. Warranties also provide for a remedy when this does not occur. Typically, they require the contractor to return to a facility and fix or replace any item that is not functioning to the requirements of the contract documents. The contractor typically backs up the warranty with warranties from equipment suppliers. When a potential warranty issue has been identified, it is essential that facts be documented carefully to avoid acting on incorrect assumptions about who is responsible for a given situation.

9.0 SAFETY CONTROL METHODS AND PROCEDURES

The Occupational Safety and Health Act (OSHA) established a nationwide safety program for workplace safety and health (OSHA, 1970). Employers are required to promote a work environment free from recognized health and safety hazards and to meet OSHA requirements (OSHA, 2006). Construction contracts contain articles that require the contractor to conform to all applicable laws, ordinances, rules, and regulations related to safety. The contractor also must comply with any safety program policies instituted by public and local governments. Additional information on safety programs and policies related to wastewater collection systems is provided in Chapter 9.

10.0 UTILITY DAMAGE CONTROL (ONE-CALL SYSTEMS)

Underground utilities pose significant risks during construction because information about their location may be lacking. In 1996, the U.S. Department of Transportation Office of Pipeline Safety organized the Damage Prevention Quality Action Team to develop a national damage prevention campaign now known as Dig Safely. Each state has set up what is referred to as a one-call, or dig-safe system, so that one phone call locates all recorded utilities in the area of construction and notifies appropriate utility companies (http://www.digsafely.com/contacts.htm). The APWA has developed *Recommended Marking Guidelines for Underground Utilities* (APWA, 2001). Uniform color-coded cards are available to standardize marking locations, changes in direction, and dead ends of buried lines. Proper utility location helps avoid accidental damage and service interruptions.

Subsurface utility engineering is the process of locating and depicting the location of underground utilities according to an industry-established level of quality. This practice has been promoted by the Federal Highway Administration as a cost-effective means of providing accurate underground utility information for design, which otherwise may cause costly construction delays.

11.0 ALLIANCES AND PARTNERING

Owners have become frustrated with disputes and the resultant effects on cost and schedules. Partnering has evolved as a concept to help head off disputes before they become an issue. Partnering is a method to open lines of communication and improve the process.

The CII (1991) has defined partnering as: "A long-term commitment between two or more organizations for the purpose of achieving specific business objectives by maximizing the effectiveness of each participant's resources. This requires changing traditional relationships to a shared culture without regard to organizational boundaries. The relationship is based on trust, dedication to common goals, and an understanding of each other's individual expectations and values. The expected benefits include improved efficiency and cost effectiveness, increased opportunity for innovation, and the continuous improvement of quality products and services."

The partnering process involves a partnering development meeting, a joint partnering workshop, periodic evaluations, required escalation of an issue, and final evaluation and project commemoration.

12.0 REFERENCES

American Institute of Architects (2006) Design Build Matrix. http://www.aia.org/SiteObjects/files/50_StateMatrix_DesignBuild_2006.pdf (accessed May 2008).

American Public Works Association (2001) *Recommended Marking Guidelines for Underground Utilities;* American Public Works Association: Chicago, Illinois.

Construction Industry Institute (1991) *In Search of Partnering Excellence,* SP17-1; Construction Industry Institute, the University of Texas at Austin: Austin, Texas.

Construction Industry Institute Dispute Prevention and Resolution Research Team (1995) *Dispute Prevention and Resolution Techniques in the Construction Industry,* Research Summary 23-1; Construction Industry Institute, the University of Texas at Austin: Austin, Texas.

General Building Contractors of New York State (1972) *General Building Contractors of New York State v. the City of Syracuse.* http://www.gbcnys.agc.org/public/owners/Single Responsibility.asp (accessed May 2008).

Knoll, J. L. (1997) *The Construction Law Briefing Paper, 97-02: Lovering-Johnson, Inc v. City of Prior Lake,* 558 N.W. 2d 499 (Minn. Ct. App. 1997), Fabyanske, Westra & Hart, P.A.: Minneapolis, Minnesota.

Kentucky Legislative Research Commission (2007) *Kentucky Revised Statutes,* Title XXVI, Occupations and Professions, Chapter 322, §322.360. http://www.lrc.ky.gov/ KRS/322-00/060.PDF (accessed May 2008).

New York State (2008) *AKA, Wick Law,* NYS General Municipal Law, Article 5-A, Public Contracts, §101. http://public.leginfo.state.ny.us/menugetf.cgi (accessed June 2008).

New York State Education Department. *New York State Education Law,* NYS Education Law, Article 145, §7209, Special Provisions. www.op.nysed.gov/article145.htm (accessed August 2007).

Occupational Safety and Health Administration (1970) *Occupation Safety and Health Act,* Public Law 91-596. http://www.osha.gov/pls/oshaweb/owadisp.show_document?p_id=2743&p_table=OSHACT (accessed August 2008).

Occupational Safety and Health Administration (2006)*Occupational Safety and Health Act* (Part 1926–Safety and Health Regulation for Construction) http://www.osha.gov/pls/oshaweb/owastand.display_standard_group?p_toc_level=1&p_part_number=1926 (accessed August 2008).

Texas Board of Professional Engineers (2008) *Texas Engineering Practice Act,* Texas Occupations Code, Title 6, Chapter 1001, §1001.407 and §1001.053, Public Works. www.tbpe.state.tx.us/eng_req.htm (accessed August 2007).

U.S. Environmental Protection Agency (1987) *The Water Pollution Control Revolving Fund* (Amendments to the Clean Water Act). http://www.epa.gov/owm/cwfinance/cwsrf/index.htm (accessed August 2007).

U.S. Environmental Protection Agency *Wetlands (Section 404) Compliance Monitoring* http://www.epa.gov/compliance/monitoring/programs/cwa/wetlands.html (accessed August 2007).

U.S. Environmental Protection Agency (2008) *National Pollutant Discharge Elimination System (NPDES) Stormwater Program's Construction General Permit.* http://cfpub.epa.gov/npdes/stormwater/authorizationstatus.cfm (accessed December 2008).

13.0 SUGGESTED READINGS

American Public Works Association (1977) *How to Form Utility Location and Coordination Committees;* American Public Works Association: Chicago, Illinois.

American Public Works Association (1978) *One-Call Systems Manual;* American Public Works Association: Chicago, Illinois.

American Society of Civil Engineers (1991) *Avoiding and Resolving Disputes During Construction;* American Society of Civil Engineers: New York.

Barrie, D. S.; Paulson, B. C. (1992) *Professional Construction Management Including CM, Design–Construct and General Contracting,* 3rd ed.; McGraw-Hill: New York.

Clough, R. H.; Sears, G. A.; Sears, S. K. (2005) *Construction Contracting: A Practical Guide to Company Management,* 7th ed.; Wiley & Sons: New York.

Clyde, J. E. (1983) *Construction Inspection: A Field Guide to Practice;* Wiley & Sons: New York.

Construction Specifications Institute (2004) *The Project Resource Manual (PRM): CSI Manual of Practice,* 5th ed.; Construction Specifications Institute: Alexandria, Virginia.

Common Ground Alliance Dig Safely One Call Contacts. http://www.digsafely.com/contacts.htm (accessed September 2008).

Del Re, R.; McKittrick, H. V. (1985) *The Rule of the Resident Engineer.* American Society of Civil Engineers: New York.

Lambert, J. D.; White, L. (1987) *Handbook of Modern Construction Law;* Prentice-Hall: Englewood Cliffs, New Jersey.

Meier, H. W. (1989) *Construction Specification Handbook,* 4th ed.; Prentice-Hall: Englewood Cliffs, New Jersey.

Chapter 7

Public Policy and Community Relations

1.0	PUBLIC POLICY THEORY 160	6.0	COMMUNITY RELATIONS AND PUBLIC EDUCATION 166
2.0	PUBLIC POLICY IN WASTEWATER COLLECTION SYSTEM MANAGEMENT 161	7.0	COMMUNITY RELATIONS AND PUBLIC MEETINGS 167
3.0	ESTABLISHING WASTEWATER COLLECTION PUBLIC POLICY 162	8.0	COMMUNITY RELATIONS AND CUSTOMER SERVICE 169
4.0	KEEPING PUBLIC POLICY RELEVANT 163	9.0	COMMUNITY RELATIONS AND EMERGENCY PREPAREDNESS 171
5.0	PUBLIC POLICY AND COMMUNITY RELATIONS 165	10.0	REFERENCES 171
		11.0	SUGGESTED READINGS 171

The goal of this chapter is to introduce wastewater collection system managers to the role of public policy and community relations in successful, sustainable system management. Every political subdivision, including city, county, state, district, or other authority, enacts laws, makes policies, and allocates resources to achieve specific end. Public policy is the process of defining problems, building support, identifying solutions, implementing policies, and evaluating effectiveness of policies.

1.0 PUBLIC POLICY THEORY

Public policy is not a single, standalone decision by a governing body but a dynamic, interactive process that reflects the common mind of the public body. Public policy formation consists of five steps as shown in Figure 7.1. The time associated with each step can vary from weeks to years, depending on complexity of the problem and agreement within the public body.

Problem definition is the first phase. In this phase, a problem is identified with corresponding goals. Often, some amount of research and study is done on the issue to develop possible concept-level solutions. Agenda setting is the second phase; although some experts group problem definition and agenda setting into a single phase. In the agenda-setting phase, efforts are made by small groups with a special interest to public raise awareness and support for the issue.

The policy-adoption phase consists of the development of possible alternatives and selection of a single solution or policy. This phase typically includes public discussions on the issue and may incorporate input from groups, such as cost estimation for the various alternatives identified.

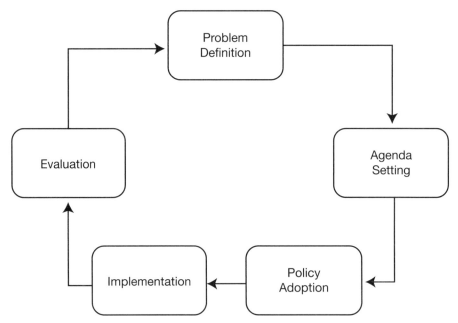

FIGURE 7.1 The five steps of the public policy cycle.

The implementation phase puts the policy into action. Further decisions are made as theoretical approaches meet real-world situations. Those responsible for carrying out policy enact procedures. Depending on the nature of the policy, the implementation phase can include testing of the validity of the policy, such as through litigation.

The final phase is evaluation. After implementation is complete, the policy is reviewed to assess the degree to which the goals of problem definition and agenda setting were met. With gaps identified, the process begins again.

The issues that reach a governing body reflect the values and priorities of the community. Typically, scientific research alone does not play a significant role in agenda setting. Instead, the issues and causes that are embraced by the public are the topics that find their way through the public policy process. As such, perception of a situation can carry as much weight or more than the reality of a situation.

2.0 PUBLIC POLICY IN WASTEWATER COLLECTION SYSTEM MANAGEMENT

Public policies affecting wastewater collection system management exist at every level of government. At the federal level, the Clean Water Act (CWA) first set the goals for clean water in 1972. Through the U.S. Environmental Protection Agency, policies were developed to affect those goals throughout the United States. Following the public policy cycle, the CWA has been amended several times to reflect the changing needs of society.

At the state level, additional regulations have been enacted to set goals to protect valued environmental assets. Although these regulations are coordinated with federal regulations, they frequently go further to provide additional or targeted protection. Similarly, regulations and policies are developed at the local level to further protect a locally valued environmental feature.

The purpose of defining all levels of public policy for wastewater collection system management is to set the baseline for characterizing current conditions, forecast those of the future, and manage the period in between. For any given condition, government can choose to intervene or to take no action. The decision of whether to intervene and how needs to be based on current conditions, such as land use, water use, and habitat, and how those conditions are likely to change. Because the environment is a dynamic, complex system, even small actions can have an effect, including unexpected, if not unwanted, side effects.

Characterizing the conditions that exist today can be done in many ways, ranging from occurrence documentation to technical analysis. The occurrence and effect of an ongoing problem can be documented by property owners, special interest groups, or wastewater collection system personnel using photographs and testimonials as evidence. Characterization using technical analysis may be done by in-house wastewater collection system staff or by a consultant. This analysis could include data collection, such as flow monitoring and water quality sampling, laboratory analysis, or the development and use of mathematical modeling tools, or some combination of these.

There are also many methods available for forecasting future conditions. Selection of a forecasting method should consider that environmental forecasting is often long-term; subsequent changes in condition are likely. It should also be considered that environmental trends and action can interact to create new concerns. For example, correcting a manhole overflow may result in a basement backup. Wastewater collection, treatment, and discharge are one small segment of environmental management. Forecasting the effects of decision making can prove challenging when actions affect segments of the environment outside the normal limits of the wastewater industry.

3.0 ESTABLISHING WASTEWATER COLLECTION PUBLIC POLICY

As described earlier, in each municipality, county, district, or authority, the public puts forth issues and priorities that need to be diligently considered and incorporated into the way funds, people, and equipment are allocated. The methods that create public policy are unique to each community, reflecting individual histories and cultures. However, within the typical realm of a wastewater collection system, public policy topics include

- Pollution of water bodies from the discharge of untreated wastewater,
- Property damage from basement or structure flooding,
- Nuisances, such as rats and cockroaches,
- Odors,
- Streambed erosion,
- Water reclamation and reuse,

- Private lateral maintenance, and
- Design and construction of new and upgraded facilities.

Other community-wide policy issues commonly affecting collection systems management decisions include affordability of rates and financing, asset investment, service extension, public-private partnerships, and economic development.

The wastewater industry often does not invest the time or resources to educate constituents or customers about the shared roles and responsibilities in managing use of the environment. As a result, the public may have performance expectations of the wastewater collector that are not possible or that are unrealistic based on the rates paid and the size of the wastewater collection system department.

Establishing local public policy recognizes and respects that the problems of the ratepayer are a problem of the community. For some issues, the policies established by the wastewater utility largely can address the problems. This is especially true when the wastewater utility owns and operates the system directly linked to the problem. For example, where odor is a problem around a pumping station, the wastewater utility has the facilities, personnel, and knowledge to install odor-control units or change the conditions to alleviate odor generation.

The utility also can establish policies that set boundaries between the responsibilities of the utility and an adjacent entity, industry, or ratepayer. Basement flooding problems frequently fall into this category. Many wastewater utilities have limited the involvement of their personnel in flooding incidents to publicly owned portions of the system. Investigation of sewer laterals on private property is the responsibility of the homeowner. Establishment of such policies is often made for practical reasons, such as limitations of staff or equipment resources.

4.0 KEEPING PUBLIC POLICY RELEVANT

The key to keeping public policy relevant is to review the policy and update it when appropriate. The day-to-day responsibilities of the wastewater collection manager typically leave little time to review policy. Yet this review is an important part of having a collection system that is viewed by utility staff and ratepayers as successful. Policy reviews should be periodic but spaced far enough apart to determine if it is achieving desired goals and, if not, where adjustments are needed. A policy may have multiple review periods that should be specific to the policy.

For example, a wastewater collection utility establishes an asset-management policy that does not include catastrophic infrastructure failures. One implementation goal is to perform small maintenance repairs to avoid large system failures. To meet this goal, the manager chooses to establish a program to televise the entire collection system once every 3 years. Under this example, the following may be an appropriate review cycle for the program:

- Every six months—review progress to determine if the program of cleaning the entire system every three years is on schedule. Review staff hours and equipment usage.
- Annually—as part of budgeting for next fiscal year, review operations and maintenance (O & M) needs to maintain the program and recommendations for capital repair projects, such as lining. Review point repairs made and project funds needed for next year.
- After three years—upon completion of the first cycle, review use of staff and equipment, point repairs made, capital repair projects, and any emergency repairs or failures that occurred. Assess how well the procedure achieved the goal and make any adjustments. Assess whether the goal continues to be relevant. To the extent possible, coordinate with other actions under way to meet the policy goal, such as installation of capital assets.
- Years four and five—review results of inspections for the two periods. Assess whether a three-year period is appropriate, too short, or too long for individual sections of the system. Is the system condition better or worse than in the first cycle? If worse, are there specific causes of deterioration not understood or known when the policy implementation program was put in place?
- After six years—same as after three years; assess system-wide whether the three-year period is appropriate, too short, or too long. Assess how well the policy is achieving the goal. Assess whether the goal continues to be relevant.

Although it can be challenging to make the time to review policies, the evaluation phase is critical to achieving the goal. Review periods should be tied to an event such as a specific a maintenance activity, completing an inspection cycle, or a management activity, such as budgeting. Although significant time and effort are expended developing policies, there is no reason to expect that the first attempt at addressing a problem will find the perfect solution. The evaluation phase provides opportunity to make adjustments.

One component of the evaluation phase is to assess the goal itself and to determine if it is still relevant. A goal may be removed from public policy if (1) it is met and is no longer needed or (2) if the public no longer values the goal and is not willing to support it. In general, goals that are part of environmental decision making are goals that, once achieved, require some amount of continued, ongoing attention. An example of this is the significant effort under way to clean and protect the Chesapeake Bay. Once the Bay has rebounded to the point desired, a policy will need to remain in place to protect the Chesapeake from degradation by future use.

The political will to retain a goal and public policy may wane if the public no longer sees a condition as a problem. Because public policy is rooted in the concerns of the public, it is subject, to some extent, to the cyclical nature of society. The wastewater collection manager can affect the sustainability of a policy by supporting sound policies and allowing others to fade.

5.0 PUBLIC POLICY AND COMMUNITY RELATIONS

Community relations are the link between ratepayers (the public) and policymakers. Through community relations, information moves both from the public to the policymakers and back from the policymakers to the public. Effective community relations are important to ensuring that ratepayers view the wastewater collection system as successful. The wastewater collection manager serves as the link between the public and policymakers. Community relations are a broad term that is used here to mean the system by which the wastewater collection system and ratepayers communicate. Terms such as public relations and public outreach are synonymous.

Each wastewater collection system has a way for ratepayers to indicate problems, express concerns, and share experiences. Frequently, this is limited to a customer service line that ratepayers can call when they have a problem. A few collection systems use a more proactive approach that strives to communicate with people when they are not in crisis, including the use of focus groups and community advisory teams. Working in community relations can require a lot of energy and a positive attitude. It is not a job everyone can do, but all staff should understand that they are the daily face of the organization. Ratepayers' perception of the utility will come from their experiences with the person on the phone and in the street—not the wastewater collection system manager.

As with technical goals, policies should be established based on managerial goals and address how the utility will interact with the public. The public is customer of the wastewater collection system, so it is important to understand their needs. Tools such as performance evaluations and surveys may provide valuable insight to their priorities and perceptions and enable a utility to better tailor services.

6.0 COMMUNITY RELATIONS AND PUBLIC EDUCATION

Public education is a facet of community relations through which information about the wastewater utility is disseminated. The purpose typically is to inform the public about activities in the utility but also can include soliciting input on specific topics. There are many formats available for public education:

- Mail:
  - Bill stuffers,
  - Newsletters, mailings, and
  - Flyers, handouts, posters;
- Media:
  - Web sites,
  - Newspaper articles, and
  - Television or radio news stories;
- E-mail: and
- One-to-one:
  - School presentations,
  - Booths at public events, and
  - Public meetings.

The Water Environment Federation's (WEF's) *Survival Guide: Public Communications for Water Professionals* (2002) assists the wastewater collection manager in developing a communication strategy and tools and in working with the public, interest groups, and the media. The Water Environment Federation also has many brochures, bill stuffers, posters, and other public education materials available through its Web site http://www.wef.org.

Selecting a public education method should be based on (1) who the target audience is; (2) how they can most effectively be reached; and (3) what results are desired. For example, if a utility wants to educate people on a new interceptor project, the target audience may be identified as ratepayers in an effort to garner their support. In this situation, bill stuffers, newspaper articles, and participation at public events may be appropriate choices, whereas school presentations and flyers in apartment complexes likely will miss the target audience. The public education methods selected should also consider the demographics of the public. For example, print material using a larger font size will make reading easier for the elderly; non-English language copies could be available in select areas.

It is often said that wastewater is an "out of sight, out of mind" industry, but that everyone wants to ensure that they are getting the most for their dollars. Public education is the way a utility can promote accomplishments and explain problem-solving approaches.

7.0 COMMUNITY RELATIONS AND PUBLIC MEETINGS

The public meeting is another way to convey information on topics of critical concern. Of the public education methods listed in the previous section, special attention should be paid to public meetings because they provide an opportunity for face-to-face interaction between put the wastewater collection system manager and the public. If not well-planned, however, public meetings can create more issues than they resolve.

Several guides and Web sites are available to help wastewater collection system managers, some of which are listed at the end of this chapter. The U.S. Fish and Wildlife Service, U.S. Department of Agriculture, Forest Service, and Oregon Department of Fish and Wildlife put together *The Public Meeting Survival Guide* (1990), which contains excellent advice on how to structure and prepare for public meetings. The following sections are based on information provided in this guide.

A public meeting may be appropriate when seeking information, advice, and involvement in making decisions, clarification on an issue, or when multiple groups are involved. The public will interpret a "meeting" to mean that the utility is willing and able to accept input. In some situations alternate public education methods might be preferred, such as if a decision has already been made, there is inadequate data, there is not enough time to manage the input, or if emotions are too high to allow constructive input. In certain situations where a decision has been made, however, it may be beneficial for the manager to hear stakeholder concerns and to explain why the work is needed or why the work needs to be performed as planned.

Once it has been decided that a pubic meeting will be held, the next question is how much involvement is desired or expected. A "presentation" or other public education method, such as newsletter or newspaper article, may be a better forum for sharing a decision while limiting involvement or input. Public meetings are interactive and allow all attendees to participate. They are informal and produce a general record of events. For some situations, a public hearing may be required. Hearings are formal and must meet legal requirements. Formal statements are sought and a verbatim record is generated.

In planning the public meeting, the manager should consider who is the audience. Questions to be considered include:

- Who are they?
- What do they know about the topic?
- What is their attitude about the topic, key stakeholders, and the wastewater collection utility?
- Are all of the side issues represented?
- Are all affected groups represented?
- Who are the decision makers and will they be present?

Consider what type of meeting format will suit the goals and audience. Examples of meeting formats include open houses, small-group breakouts, panel discussions, question and answer, presentation and discussion, and explanation and testimony. Small-group breakouts are excellent for giving everyone an opportunity to voice opinions. One drawback is that two people may have a different experience based on who led the small group and comments were managed. Care should be taken in all formats to remain compassionate and open but not try to solve everyone's problems on the spot, which can be interpreted as dismissing the concern.

Time needs to be taken to properly plan the meeting. This process begins by identifying meeting objectives. One way to do this is to start a sentence with: "at the end of this meeting, I want the audience to." This sentence can help guide the rest of the planning. For public meetings, objectives should be sensible, measurable, attainable, relevant, and trackable. Managers should keep several suggestions in mind:

- Create an agenda that is followed;
- Ensure meetings begin on time;
- Include breaks to keep the audience comfortable and focused;
- Alternate speakers and activities every 10 to 15 minutes, if possible;
- Select methods for receiving input—brainstorm lists, written comments, verbal feedback, prioritized preferences, questionnaire, or email; and
- Provide information to take away, such as handouts or flyers.

Slideshows are a common and effective media to use during public meetings. When developing slides, managers should consider the following guidelines:

- Set up and test equipment before the meeting begins;
- Include a maximum of 4 to 5 information items per slide;
- Use animation selectively; too much can be distracting for the audience;
- Use pictures and graphics to demonstrate points; and
- Discourage speakers from reading slides to the audience; rather, they should elaborate on information provided.

Also, speakers should avoid industry jargon, especially acronyms, that are not common within mainstream society. If it cannot be avoided, terms will need to be defined for the audience throughout the presentation. Otherwise, the presentation will sound as if it is in a foreign language and the information that being conveyed will be lost.

Managers should be aware of any commitments made during the meeting to ensure follow through. For example, if a utility representative promised that meeting minutes would be available in 24 hours and they were not, credibility will suffer. As an alternative, a timetable without exact dates can be provided with ways for people to submit follow-up comments or ideas.

It is important that utility representatives be honest and compassionate in public meetings to demonstrate respect and value for the role of everyone in the room.

8.0 COMMUNITY RELATIONS AND CUSTOMER SERVICE

For this discussion, customer service is the direct interaction between ratepayers and the staff of the wastewater collection utility. Commonly, customer service personnel manage day-to-day activities such as activating new accounts, closing accounts, accepting payments, answering questions, and responding to requests for service. Typically, customer service activities are performed by people who are dedicated to customer services but, especially in larger organizations, do not work with the engineers and operators who manage the system. Internal policies must be established to set a level of service that each customer can expect, such as targeted maximum waits hold and response times.

A request for service is passed from the phone operator to the collection system operator for action. The collection system operators are in the challenging position of responding to a situation that has caused the customer some inconvenience or damage, such as basement flooding. Customers can be emotional—angry, distraught, confused—and expect the operator to help improve the situation. Although there are times when the operator can help, they frequently must tell customers that the problem is on private property and explain that policy limits what they can do. Collection system operators are trained in how to inspect, operate, and maintain a collection system, but typically not in how to interact with a person under duress. Many operators find this to be an especially burdensome and stressful part of the jobs. Significant time can be required to talk with the customer, affecting the overall productivity of the crew. Some communities have found success designating a person or persons to interact directly with the public about field work. This person allows crews to focus on technical performance of the job and provides the time and detail the customer needs. Whether this aspect of customer service is provided by the crew or a designated person, wastewater collection system managers must provide the interpersonal communications skills training needed to effectively perform the job. Poor customer service can make a difficult situation worse, taking up more staff time and resources. Providing good customer service will ease the situation for customers and, in turn, improve the utility's standing within the community.

Internal policies also must be established for the performance of service request work. Policies for service requests aim to provide a high level of customer service and typically are based on the performance of work and maintenance of equipment. Common policies include initial response to a service request within 24 hours, closing out of a service request within seven days, issuing service performance questionnaires, and calling to follow up on the service and equipment repairs complete within 10 days.

Customer service calls can be an indicator of the priorities of the public, which feed into the development of public policy. An electronic database should record calls, including the address of the caller, description of the complaint, and date of call. This provides a searchable data record that can be used to examine which parts of the service area have the most problems, whether problems are associated with a specific time of year, and what types of problems are prevalent. After the service request is resolved, this dataset can be expanded to include cause of the problem, action taken, and man and equipment hours used. The collection and use of this type of data is discussed further in Chapter 3, Information Management.

9.0 COMMUNITY RELATIONS AND EMERGENCY PREPAREDNESS

In the event of an emergency, community relations will be the channels first tapped to convey information to the public. Since Hurricane Katrina and the disaster of the World Trade Center attack in New York on September 11, 2001, the nation has developed emergency preparedness guidelines to disseminate information and share resources (see Chapter 10). On a community-wide basis, the police or fire departments, or both, typically manage emergencies, often using a media spokesperson to speak to the public and press. However, in the event of a wastewater collection system specific emergency, such as a sewer collapse, the responsibility of communicating the extent of the situation to the public or to the police and fire departments may reside within the wastewater collection utility. Utilities should have in place plans for communications with customers and stakeholders during disaster conditions. This "crisis communication plan" should include drafts of messages that may be needed in a disaster. Wastewater collection managers are strongly encouraged to review the emergency management system in place in their community or region and consider how the utility can tap into the system in the event of an emergency.

10.0 REFERENCES

U.S. Fish and Wildlife Service; U.S. Department of Agriculture; Forest Service; Oregon Department of Fish and Wildlife (1990) *The Public Meeting Survival Guide*; U.S. Fish and Wildlife Service; U.S. Department of Agriculture; Forest Service; Oregon Department of Fish and Wildlife: Portland, Oregon.

Water Environment Federation (2002) *Survival Guide: Public Communications for Water Professionals*; Wantland, S., Ed.; Water Environment Federation: Alexandria, Virginia.

11.0 SUGGESTED READINGS

Armstrong, J. S. (1999) Forecasting for Environmental Decision Making. In *Tools to Aid Environmental Decision Making*; Springer-Verlag: New York.

Barkenbus, J. (1998) *Expertise and the Policy Cycle*; Energy, Environment and Resources Center, University of Tennessee: Knoxville, Tennessee.

Dolan, R. J.; Rose, T. D.; Baker, R. A.; Barnes, M. J. (2004) *Managing the Water and Wastewater Utility*; Water Environment Federation: Alexandria, Virginia.

Innes, J. E.; Booher, D. E. (1999) Consensus Building and Complex Adaptive Systems. *J. Am. Plan. Assoc.*, **65** (4), 412.

International Association for Public Participation Home Page. http://www.iap2.org (accessed November 2008).

Karkkainen, B. C.; Fung, A.; Sabel, C. F. (2000) After Backyard Environmentalism. In *The American Behavioral Scientist*; December, p 692.

North Carolina Grantmakers (2005) How to Engage In Public Policy. *Public Policy Grantmaking Toolkit.* www.ncg.org/toolkit/html/gettingstarted/howtoengage/print_howtoengage.html (accessed November 2008).

North Carolina Grantmakers (2005) Definitions. *Public Policy Grantmaking Toolkit* www.ncg.org/toolkit/html/gettingstarted/definitions/print_definitions.html (accessed November 2008).

Chapter 8

Budgeting and Financial Planning

1.0	INTRODUCTION 174	5.1	Capital Funding Sources 188
2.0	OPERATIONS BUDGET 175		*5.1.1 User Charges* 188
3.0	CAPITAL IMPROVEMENT BUDGET 179		*5.1.2 Ad Valorem Taxes* 189
4.0	SETTING BUDGET PRIORITIES THROUGH ASSET MANAGEMENT 181		*5.1.3 Special Assessments* 189
			5.1.4 Impact Fees 189
	4.1 The Problem 182		*5.1.5 Grants* 000
	4.2 The Solution 182	5.2	Capital Financing Sources 190
	4.3 Underground Infrastructure: The Tough Asset 183		*5.2.1 General Obligation Bonds* 190
	4.4 Life-Cycle Cost Analysis 185		*5.2.2 Revenue Bonds* 190
	4.4.1 Equivalent Annual Cost Approach 186		*5.2.3 Governmental Loans* 191
	4.4.2 Common-Life Approach 186	6.0	REVENUE REQUIREMENTS AND FEE SETTING 191
	4.4.3 Steps to Completing a Life-Cycle Cost Analysis 187	6.1	Financial Condition 192
		6.2	Establishing the Basis for Fees and Charges 193
5.0	CAPITAL FUNDING AND FINANCING OPTIONS 188		*6.2.1 Cash-Needs Approach* 193
			6.2.2 Utility Approach 194

(continued)

6.2.3	Selecting a Test Year	195	7.0 ROLE OF GOVERNMENT ACCOUNTING STANDARDS BOARD STATEMENT NUMBER 34 IN BUDGETING AND FINANCIAL PLANNING 198
6.2.4	User Charge Revenue Requirements	195	
6.3	User Charge and Service Fee Types	196	8.0 SELLING THE ASSET-MANAGEMENT APPROACH AND BUDGET PLAN 200
6.3.1	User Charges	196	
6.3.2	Impact Fees	197	
6.3.3	Other Fees	198	9.0 REFERENCES 202

1.0 INTRODUCTION

Budgeting and financial planning are important aspects of wastewater collection system management. Prudent financial management ensures that sufficient revenues are available to fund all necessary operations and maintenance (O & M), repair, and replacement programs. Financial management is necessary to maintain collection systems in good working condition, and as necessary, to generate sufficient revenues for system expansion to serve new customers.

This chapter will introduce the reader topics related to financial budgeting principles and practices:

- The operating and capital budgeting process;
- Economic analyses that can be used to determine the life-cycle costs of project alternatives;
- A recommended approach to setting budget priorities through asset management;
- Capital funding and financing alternatives commonly used to pay for wastewater collection system projects;
- Methods of determining revenue requirements and establishing user charges and service fees;
- The role of Government Accounting Standards Board, Statement Number 34 (GASB 34), in the budgeting and financial planning process; and
- Selling an asset-management approach.

2.0 OPERATIONS BUDGET

The operating budget is the primary financial plan for a utility. It guides management during the budget period (typically a fiscal year, but more frequently covering multiple years) for acquiring and applying resources to produce the targeted level of service. Development of an operating budget consists of three basic steps:

(1) Preparation of the budget in a format that allows it to be reviewed and discussed at all organizational levels.
(2) Review of the budget by approving authorities at various levels within the organization.
(3) Presentation and adoption of the budget at the highest level of authority required to invest it with legitimacy.

Preparation of the budget begins with the definition of service level to be provided. This can become time consuming as organizations attempt to go through highly theoretical approaches to defining missions, goals, objectives, and desired results. However, there is a simple straight-forward approach: Start with the level of service that is currently being provided. This has the advantage of establishing a common ground of understanding throughout the organization. Although current level of service may not be formally established or defined as clearly as theoretically desirable, problems and successes may have emerged through the operation of the system in several forums: internal discussions and problem solving sessions, employee/manager performance reviews, customer complaints and problem resolution, and meetings with top management and the governing body.

This current level of service can be implicitly accepted: "With this budget we will be able to continue at the same level during the coming year (or other budget period)." Or an attempt can be made to define what "current level" means in terms of organizational goals, objectives, and operational results. If so, these should be kept fairly simple because the budget must be clear and understandable at all levels of the organization and to the public. For example, level of service could be defined in terms of:

- Uninterrupted conveyance of wastewater flow to the treatment facility;
- Wastewater flow to the treatment facility with less than two household backups per year;
- Response time to critical incidents; or
- System condition, expressed in terms of condition evaluation efforts.

Whatever service indicators are selected, the budget process should identify the level of service that is currently being provided.

The next step in the process is to value the current level of service in terms of the resources that are necessary to maintain the wastewater collection system and ongoing costs that are incurred. The cost of owning and operating a wastewater collection system typically include the following cost components:

- Personnel services costs (including salaries, wages, and employee benefit).
- Material and supplies costs.
- Machinery and equipment (minor capital) costs.
- Debt service costs (i.e., principal and interest payments associated with outstanding debt, if any).
- Taxes (such as property, gross receipts, or income taxes for investor-owned utilities; excise tax on gross receipts or payments in lieu of taxes for municipally owned utilities).
- Transfers (e.g., transfers to the general fund for services for municipally owned utilities, if any).

To develop a budget for the current level of service, a manager simply starts with the current period budget categorized as shown above. Within these categories, there will be more detailed line-items of expenditures, normally defined in the utility's chart of accounts. Also available will be actual expenditures to date for the current period as well as actual expenditures compared to budget for prior periods. This information provides a baseline for the manager to begin budget preparation.

The primary question is: "what resources are needed to continue the current level of service?" This is generally defined as the resources within the current budget. The next step is to value those resources in monetary terms for the prospective budget period. Generally, this is a simple process: Current levels of expenditure are increased by the expected level of inflation from the current period to the prospective period. Often, the utility's finance department or budget office will provided inflationary factors. In fact, it has become common for these groups to provide to operating managers with a forecast of the current budget compared to the prospective budget period, given assumed levels of inflation. If this is not the case, reasonable estimates of inflation should be developed by the manager.

The consumer price index published by the Bureau of Labor Statistics can be used as a guide for inflating personnel services and materials and supplies expenses (www.bls.gov). Another common guide for inflating costs is the Engineering News

Record's construction cost index for a particular region. A more precise estimate might be obtained from the public service commission in the state, which often publishes an inflationary factor for use by regulated utilities. In addition, the utility's system for providing merit increases to employees should be considered and a reasonable estimate of those increases should be prepared. In an organization using collective bargaining, negotiated (or expected) labor contracts for salaries, wages, and benefits should override general inflationary factors. Also, contract agreements established with vendors for future costs should be taken into consideration when forecasting the cost of purchases of materials, supplies, and equipment.

The budgeted cost associated with paying debt service on existing, outstanding debt can be projected based on actual debt amortization schedules provided with bond and loan documents. Other costs, such as taxes and transfers, can be budgeted based on historical trends and discussions with general fund managers, agencies, or departments imposing such cost on the utility. The initial proposed budget for continuing the current level of service should, then, be composed of:

- A statement of the current service level in as precise terms as possible, including the expected output of the organization.
- An expenditure budget by line item or object of expenditure, as defined in the organization's chart of accounts, which essentially consists of the current level of expenditure increased by expected price inflation to the prospective budget period.

This current service level budget can stand alone as the organization's budget proposal. However, it also can form a baseline for alternative budgets, which fall into two general categories:

- Service improvement options, or
- Service reduction options.

To develop these alternative budgets, it is extremely important that current service level be defined clearly because the heart of the process is the ability to determine consequences on service level of alternative spending plans.

Service improvement options should be constructed to provide a clear connection between increased expenditures and improved service level. There also may be known cost increases, such as power costs for lift stations, increased fuel costs, or planned staffing increases because of new, previously approved infrastructure coming online. These types of increases may be necessary simply to maintain the current level of service and should be presented as such. If not approved, these cost increases would result in a lower level of service, which should be documented. The goal of providing a clear connection between expenditures and service level is best served by constructing these budgets in relatively small increments from the baseline (current service level), generally in the range of 3 to 10% overall increase. This allows reviewers to focus incrementally on costs and service level effects and to make an informed decision regarding suitability of the investment for the organization. Service improvement options should be constructed using the same format and line items as the baseline budget, so that they easily can be compared and evaluated.

Service reduction options are constructed exactly like service improvement options, but represent reductions in expenditures and level of service from the baseline.

Each service change should be accompanied by a clear, precise statement of the effect of the increase or decrease in expenditures on service as represented by the baseline budget. Taken together, this constitutes a system of "target budgeting" that is highly manageable and transparent at all levels of the organization that have responsibility for development, review, and approval of the budget. A column format for the final document would allow easy comparison among options, as illustrated in Table 8.1.

TABLE 8.1 Example statement of the effect on the current service level of an increase or decrease in expenditures as represented by the baseline budget.

Collection system budget options	Service reduction option 2	Service reduction option 1	Baseline (current service level)	Service improvement option 1	Service improvement option 2
Personnel costs	$8,550,000	$9,500,000	$10,000,000	$10,500,000	$10,000,000
Materials and supplies	$3,420,000	$3,800,000	$4,000,000	$4,500,000	$4,750,000
Equipment	$500,000	$950,000	$1,000,000	$1,350,000	$2,500,000
Debt service	$2,000,000	$2,000,000	2,000,000	$2,250,000	$2,500,000
Taxes and transfers	$1,500,000	$1,500,000	$1,500,000	$1,500,000	$1,500,000
Total	$15,970,000	$17,750,000	$18,500,000	$20,100,000	$21,250,000
Change from baseline	–13.7%	–4.1%	0.0%	8.6%	14.9%

Description of service level options
Service reduction 2   Elimination of proactive inspection system implemented two years ago
Service reduction 1   Reduction of proactive inspections by 40%
Baseline   5% increase over prior year budget due to inflation and merit increases
Service improvement 1 Escalation of proactive inspection system to complete in five rather than seven years
Service improvement 2 Improvements to lift stations, including telemetry

3.0 CAPITAL IMPROVEMENT BUDGET

Utilities are capital-intensive by nature, and the development of clear, realistic plans for investment in capital assets is an essential part of utility management. Capital budgets, which establish planned capital expenditures for a prospective budget period, should be based on sound long-term capital improvement programs (CIPs). In financial accounting terms, capital expenditures are distinguished from operating expenditures in that they are capitalized: the cost of a capital expenditure is entered into the accounting records as a fixed asset, then that cost is recognized over the estimated life of the asset through depreciation expense. This allows the cost of capital assets to be amortized over the life of the asset. Although this approach meets the objectives of financial accounting, budgeting must take into consideration the actual flow of financial resources needed to implement the capital program. Capital projects can be funded with cash generated through revenues of the system, or by issuing debt and paying principal and interest of the outstanding balance over a period of time consistent with the expected life of the assets. Therefore, construction and acquisition of capital assets creates the need for cash flow either to directly fund the projects or to pay the associated debt service. Cash flow requirements are the primary consideration in developing a capital budget.

This section addresses types of capital improvements, expenses, tools used to evaluate the cost and benefits of capital projects, and financing and funding of capital projects. A more comprehensive description of the capital improvement planning process can be found in Chapter 4.

A wastewater collection system capital budget should reflect two general classes of expenditures:

(1) Renewal and replacement-costs that are incurred to extend the life of current assets through the rehabilitation or replacement of the existing asset.
(2) Expansion or growth-costs that are incurred to expand the capacity of the system and meet the needs of new customers.

In addition, regulatory requirements can drive the need for some capital improvements. This need can be part of either replacement or expansion projects. For example, reduction of combined sewer overflows (CSOs) for water quality improvement may require replacement of an existing combined sewer transmission line. The replacement project may be sized to meet future demands, which means it will also have an expansion or growth element included. Identifying the purpose and drivers of capital projects is important in determining the financing options as described later in this chapter.

Capital expenditures should be categorized by class of expenditure, as noted above, and in terms of their function in the system. There are two major functional areas in a wastewater collection system:

(1) Collection—small-diameter collector pipeline directly connected to user facilities and transmitting wastewater to the transmission system.
(2) Transmission—large-diameter transmission pipelines and relating pumping facilities used to transmit wastewater from the collection system to the treatment plant.

A sound capital improvement program will be based on projections of future demand, regulatory requirements, asset system evaluation of useful life, condition assessments, and other factors that might influence the need for renewal, replacement, and expansion of facilities.

In composing a capital budget, a manager should consider the cost and benefits of capital project alternatives. In addition, the establishment of a capital budget should also refer to the long-term CIP. It is important to consider such questions as:

- Have the factors and assumptions driving the CIP changed significantly since it was developed?
- Have events been accelerated or delayed requiring delay or expedition of projects in the CIP?
- If the CIP is based on specific funding sources, are those sources still available?
- Has the political climate changed significantly so that changes are required to the CIP?

This part of the process requires a review of the CIP for the short term. Long-term CIPs are developed based on estimates about the future that almost always require some degree of adjustment over time. Simply incorporating the CIP for a certain year into a short-term budget plan—for which funds will actually be appropriated through a legal process—could result in significant misapplication of funds.

Once the CIP has been updated to reflect current conditions, the capital budget for the prospective budget period is ready to be formulated. First, projects should be categorized to reflect the status of the projects in the budget:

- Projects continued from prior years and
- New projects.

A budget worksheet should be developed for each project, including at a minimum the following items:

- Project name and unique identification number;

- Expected initiation and completion date;
- Type of project or percent of type (renewal and replacement, expansion, or regulatory);
- Description of purpose and advantages of project;
- Effects if project is not completed should be quantified;
- Funds acquired in prior periods that are carried forward to fund the project, if any;
- Source of new funds to supplement existing funds to the extent necessary to fund the entire project; and
- Expected effect on the operating budget of the project, showing the amount of each year for at least five years.

Based on the project worksheets, a capital budget should be composed showing each project on a separate line, clearly identifying the type, funding sources, and priority. For the prospective budget period, the total of all project expenditures and funding sources summarized, showing the total operating and capital budgets for the period.

Capital funding sources are discussed later in this chapter. If a project is debt-financed, however, the amount of cash required to repay the principal and interest and to meet debt service coverage requirements during the prospective budget period should be reflected in the cost summary for the budget.

As with an operating budget, it is important to tie capital expenditures to the utility's level of service. A summary of advantages and disadvantages related to the level of service can be developed. The next section, which focuses on asset management, explains how to determine capital needs using prioritization that is based on a clear understanding of the consequences of a capital spending plan.

4.0 SETTING BUDGET PRIORITIES THROUGH ASSET MANAGEMENT

This section addresses strategies for determining the capital budget and forecast amounts for renewal and rehabilitation of wastewater collection system assets. A more comprehensive description of the CIP process can be found in Chapter 4. Although some rules for reinvestment are referenced—such as 2% of the asset value assuming a 50-year service life or funding a minimum of annual depreciation expense for renewal and replacement projects—other, more focused approaches exist, as discussed below.

4.1 THE PROBLEM

Wastewater utility managers often speak of being stuck "between a rock and a hard place." They have to address issues such as CSOs; sanitary sewer overflows; capacity, management, operations, and maintenance; and other state and federal requirements. Managers also are challenged with "build out," "no growth," "urban decay," and "price resistance." If a utility carefully and equitably funds and finances the cost of expanding system capacity, meeting growing demand is self-balancing as rate revenues and connection (capacity) charges flow in from new customers. However, the emphasis has shifted for many systems to rehabilitation and renewal of existing assets to maintain service levels for current customers and to remediate deficiencies under regulatory requirements. Utility managers will be faced with an increasingly difficult job to secure needed rate increases as utility infrastructure ages and replacement needs increase. Although this problem is not new, the necessity for action becomes more clearly defined with every passing year. This is because the easier, softer way (replacement deferral) has too often been taken to avoid the immediate rate effect. Renewal and replacement costs are increasing.

4.2 The Solution

These problems are not limited to older systems or to economically stagnant areas. It is simply a matter of time: if this is not happening, it will happen in the future unless action is taken. In the industry, the set of methods, concepts, techniques, and systems developed to deal with these issues often are collectively referred to as "asset management". Assets can be viewed as resources that are used over their useful life to provide services and produce revenue. Good asset management involves the acquisition, coordination, and application of all assets. A good asset-management system is characterized by

- Reasonably complete inventory of all system assets,
- Reasonably accurate condition assessment of all system assets,
- Estimated useful lives for all assets,
- Existing asset values,
- Replacement costs for assets,
- Projections of total replacement costs in future years, and
- Sustainable service level.

4.3 Underground Infrastructure: The Tough Asset

Underground infrastructure is only one several assets required to run a successful wastewater utility system. Infrastructure also includes treatment plants, disposal systems, pumping stations, instrumentation, and transmission, collection, and distribution pipelines. However, buried infrastructure is out of sight, making it difficult to conduct condition assessments without considerable time and money and easy to overlook, which can result in neglect. Renewal and replacement of pipelines is a dirty, difficult job. It is also a thankless job because, as pointed out, it places a cost burden on the existing customers that is unmitigated by new customer revenues.

An effective asset-management process requires four determinations:

(1) Condition,
(2) Criticality,
(3) Cost, and
(4) Consequences.

The first step is assessment of condition. Without condition assessment, the utility manager does not have enough information about the system. There are three basic approaches to maintenance of these assets:

(1) Random response approach—fix it when it breaks.
(2) Systematic naïve approach—because it all has to be fixed eventually and generally lasts 50 years, dig up 2% per year and replace it.
(3) Systematic intelligent approach—invest money to assess asset condition and fix those needing repair.

Figure 8.1 shows the rate effects implied under each approach. The random response starts out good, but costs skyrocket as the system deteriorates. The systematic naïve approach, because it does not address the most failure-prone parts of the system first, produces significant rate effects in early years, and then levels out later on. The systematic intelligent approach, by focusing on assets in the worst condition first, produces a more level rate effect than the systematic naïve approach, but higher rates in the early years than the random response.

The second step, analysis of criticality, focuses the asset-management effort by determining the relative importance of each part of the system. A priority for repair and replacement can now be established by combining condition assessment (how likely is it that this asset will fail?) with criticality (if this asset fails, what will be the effect on the system in terms of liability, risk, and finances?).

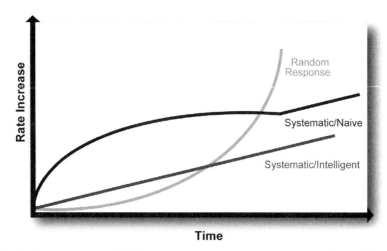

FIGURE 8.1 Rate impacts of different asset management approaches. (Courtesy of Malcolm Pirnie, Inc. All rights reserved.)

The third step, cost determination and planning, involves estimating resources required to bring the assets to the desired level of condition and scheduling repair or replacement. This results in a prioritized CIP that provides an essential cost element for the forecast of rate effects.

The final step, evaluation of consequences, provides documentation of the probable effect on the system, to service and finances, of eliminating repair or replacement projects from the plan. The consequences of funding, or not funding the CIP provides the governing body with enough information to make decisions.

The conceptual asset-management process described above requires funding through the budget process; often, the budget can be viewed as a classic sales model. Management determines what the utility needs are and communicates that to the governing body. The governing body hears from customers and balances customer needs with realities of meeting regulations and properly maintaining level of service. However, customers may exhibit price resistance ("objections"). Management provides information to answer those objections, everyone discusses the issues, and the governing body makes a decision.

In a classic arms-length sales transaction; however, either party can simply walk away, which is not an option in this situation. Resolution must be reached regardless of whether the final decision produces a budget that meets the utility's needs. This makes development of sufficient information about the utility's needs, costs, and rate effect and persuasive presentation of this information essential to effective utility management.

4.4 Life-Cycle Cost Analysis

Life-cycle cost analysis is a useful tool that can aid in development of a capital budget. Life-cycle cost analysis can be used to help determine the total cost of implementing a capital project, compare costs of project alternatives, and make capital investment decisions, such as choosing between repair or replacement or other projects. Comparing the life cycle costs of alternative capital investments is a way to evaluate their relative cost effectiveness and incorporate upfront capital costs, annual operating costs (or savings), and periodic replacement costs. Life-cycle cost analysis calculates the cost of an investment, including initial and annual O & M cost, over its life. It is especially useful in comparing the overall cost of projects that have different capital investment and O & M cost characteristics.

Under a typical life-cycle analysis, costs over the life of the project are converted to present value so that future costs can be compared on an equal basis. This "discounting" of future costs is done to reflect the time value of money (i.e., a dollar today is worth more than a dollar tomorrow). This is true because a dollar today can be invested and earn a return. Similarly, a dollar spent today costs more because of the opportunity lost from not investing and earning return. The formula for calculating present value of costs is presented below in Equation 8.1.

$$\text{Present Value} = \frac{\text{Future Value}}{1 + \text{Discount Factor}} \qquad (8.1)$$

When calculating the present value of future costs it is important to select an appropriate discount factor. The discount factor should represent the opportunity cost—the cost foregone by investing in the project today rather than placing the money in an interest-bearing investment of similar risk. The utility's investment earnings rate commonly is used because it is measurable and typically reflects a conservative investment policy that corresponds to the project risk. Other benchmarks, such as the Department of Interior Bureau of Reclamation's change in discount rate for water resources planning also has been used (DOI, 2007).

A significant limitation of present value calculations is that they can not be used to compare projects with different service lives. When comparing the life-cycle cost of mutually exclusive projects with significantly different service lives, it is important to make adjustments to account for their unequal lives. There are two commonly used adjustments to the life-cycle cost analysis that allows for a comparison of projects with unequal lives.

(1) Equivalent annual cost approach and
(2) Common life approach.

4.4.1 Equivalent Annual Cost Approach

Under the equivalent annual cost approach, the net present value of each project is calculated over its service life. The present value of each project is then converted into an equivalent annual payment, and the equivalent annual payments for each project are compared. The annualized cost is determined by equating the present value cost of the project and the present value of the annual payment (Brealey and Myers, 2000). This approach assumes that the project will be replaced with an equivalent project at the end of its life span.

4.4.2 Common-Life Approach

The common-life approach converts the projects that are being compared to the same life span by one of three methods: (1) replicating the shorter life project until its useful life span matches the longer life project; (2) replicating each project until a common life is found; or (3) completing the analysis over the life span of the shorter project and calculating and subtracting a residual value of the project with the larger life span from the present value cost.

Life-cycle cost analysis for wastewater collection system projects typically include the following costs and considerations: first cost, project life, O & M cost, and repair and rehabilitation costs.

- First cost. The first cost is simply the bid price for a project. This cost includes the construction estimate for labor, material, contingencies, supervision, and administration. Historical data may be used to determine an appropriate value for prebid evaluations. If a least-cost analysis is included as a basis for awarding a contract, the actual bid price would be used in the evaluation.

- Project life. Project life is the number of years that the project is anticipated to serve its intended purpose. For example, the project service life for culverts and sewers under primary and secondary roads may be between 50 and 100 years. A variety of sources are available to determine the life of wastewater collection system assets, including equipment manufacturers, public service commission allowable service lives, and empirical data from field testing.

- Operations and maintenance costs. These costs include the cost of operating and maintaining the project such as labor, materials and equipment, routine maintenance, and preventative maintenance costs.
- Repair and rehabilitation costs. Repair and rehabilitation costs may extend the life of wastewater collection system assets and can be considered in a life-cycle cost analysis.

Assumptions and estimates should be carefully documented, and the cost should be analyzed to ensure the best value.

When completing a life-cycle cost analysis, it is important to decide whether the analysis will consider cost in nominal or real terms. Costs stated in nominal terms take into account anticipated future trends because of price inflation. Life-cycle cost analyses are commonly completed in nominal terms (i.e., future costs are escalated for inflation). However, as long as inflation is handled consistently in the life-cycle cost analysis, either approach will result in the same conclusions. Regardless of which approach is taken, it is important to use a consistent approach to inflation of costs. If costs are escalated for inflation, all costs in the analysis should be escalated for inflation. In addition, an inflation-adjusted discount factor should be used (i.e., nominal discount factor). If costs are not escalated for inflation, a noninflation adjusted discount factor (i.e., real discount factor) should be used (Brealey and Myers, 2000).

4.4.3 Steps to Completing a Life-Cycle Cost Analysis

In summary, the following steps should be followed for completing a life-cycle cost analysis:

(1) Identify the objectives of the project and identify alternative ways of accomplishing the objectives.
(2) Estimate the cost of each alternative project including first costs, annual O & M costs, and anticipated repair and rehabilitation costs.
(3) Decide whether the analysis will consider costs escalated for inflation and use a consistent approach for inflation of costs. If costs are escalated for inflation, ensure that all costs in the analysis are escalated for inflation and a nominal discount factor is used. If costs are not escalated for inflation, ensure that a real discount factor is used.
(4) Estimate the life of each project alternative being considered. If repair and rehabilitation costs are anticipated, factor in the extension of the life of the asset because of the repair and rehabilitation efforts.

(5) Determine whether the project lives being compared are significantly different. If they are significantly different, determine whether the common life or the equivalent annual cost approach will be used.
(6) Calculate the life-cycle cost of each project using the net present value method, and make adjustments using the common life or equivalent annual cost approach as necessary.
(7) Compare the life-cycle cost of each project alternative.

5.0 CAPITAL FUNDING AND FINANCING OPTIONS

Once a project alternative is selected, funding and financing options should be considered. When considering project funding and financing it is important to refer to capital funding and financing policies that may provide guidelines for the utility. For example, a wastewater utility that plans to spend $1 million each year on collection system rehabilitation may be able to generate sufficient rate revenues to fund the program, rather than to use external financing, according to its financing policy. Alternatively, the policy may specify that a large, nonrecurring capital improvement, such as a pumping station, may need to be financed with bonds rather than funded on a pay-as-you-go basis with rate revenues. In addition, there also may be policies limiting the level of debt service allowed. It is important to carefully evaluate the various funding and financing alternatives within the policy framework and select an option that best meets the utility's needs and policies.

5.1 Capital Funding Sources

Capital funding sources are the sources of revenue that can be used to pay for capital projects and capital-related expenses (e.g., debt service). There are several capital funding sources that may be considered for wastewater collection system projects as summarized below.

5.1.1 User Charges

User charges are a source of funding both O & M and capital expenses. User charges may include any and all costs associated with the provision of wastewater service, including cash-funded capital and debt service (WEF, 2005). User charges can be used to fund capital projects on a pay-as-you-go basis, similar to O & M expenses, and to fund renewal and replacement projects.

5.1.2 Ad Valorem Taxes

An ad valorem tax is a tax that is based on the assessed value of the property rather than the flow or characteristics of the wastewater discharged. The advantages of the ad valorem tax are its assessment and administration simplicity, and its inherent ability to collect revenue from undeveloped properties in the service area that may eventually connect to the system (WEF, 2005). The disadvantages of using this funding source is that the assessment is based on property values and does not directly relate to wastewater flow and service characteristics. In addition, tax-exempt properties that may be connected to the wastewater collection system would be exempt from paying for wastewater service.

5.1.3 Special Assessments

Special assessments frequently are used where wastewater infrastructure is constructed for a specific purpose or for the benefit of a defined area, such as installing or replacing local sewer collection facilities (WEF, 2005). Property owners may pay a special assessment based on a uniform charge per household, linear front footage, or other methods. A referendum is often required to establish a special assessment district.

5.1.4 Impact Fees

Impact fees—otherwise known as capacity fees, connection fees, and system development charges—are one-time fees typically collected from a builder, developer, or new homeowner for new or expanded service (WEF, 2005). In growing or expanding communities, the revenues collected from impact fees can be a significant component of the revenue stream of the utility. Impact fees may be used to pay for growth-related wastewater collection system capital projects or for debt service payments for growth related projects. There are, however, many state-specific legal requirements that should be understood before using this funding source. One disadvantage of this funding source is that revenues collected are dependent upon growth occurring and can be variable. Another disadvantage is the potential delay between when the growth-related capital project is constructed and when the impact fees are collected.

5.1.5 Grants

The amount of grant funding that is available today to fund wastewater collection system projects is much less than the grant funding opportunities that existed during the U.S. Environmental Protection Agency grant program of the 1970s (WEF, 2005).

However, several federal and state grant funding programs still exist that potentially can be tapped to fund wastewater collection system projects. These include grants available through the Community Development Block Grant Technical Assistance Program sponsored by the U.S. Department of Housing and Urban Development, Water and Environmental Programs sponsored by the Rural Utilities Service of the U.S. Department of Agriculture, and Economic Development Administration Grants, among others. Grants have specific eligibility criteria and project requirements, including for funding cycles, schedules, wage rates, eligible cost, and in-kind matching and for conducting legal, financial, and administrative reviews. The effect of these requirements and conditions should be carefully evaluated to determine project eligibility and suitability. With the diminished availability of grant funds, competition is steep, and it is often difficult to qualify for funding. A project has a better chance of securing a grant if it helps solve a documented water quality problem or health hazard or if the utility is located in an economically disadvantaged area, especially if utility charges are greater than 1 to 3% of the medium household income.

5.2 Capital Financing Sources

Capital financing sources commonly are used to amortize capital project costs over time. There are several capital financing sources for wastewater collection system projects as summarized below.

5.2.1 General Obligation Bonds

General obligation bonds are municipal bonds that are backed by a pledge of the full faith and credit of the issuer and can only be issued by governments that have the authority to tax. This is the strongest pledge the government can provide; therefore, general obligation bonds generally have the lowest cost of the various financing vehicles available to local governments (AWWA, 2008). However, because many states have limits on the amount of general obligation bond debt that can be issued, wastewater collection system projects that are targeted for this type of financing must compete with other general fund projects.

5.2.2 Revenue Bonds

Revenue bonds are government bonds that are backed by the specific revenues that are pledged in the bond documents. Revenue bonds are commonly used to finance capital projects that generate revenue and that are expected to be self-supporting, such as a wastewater utility. Most revenue bonds are issued with provisions that provides assurance to bondholders that their money will be repaid. For example, it may include a rate covenant specifying that user charges will be established to generate sufficient rate revenues to meet all operating expenses, debt service, and debt service coverage requirements. Debt service coverage is defined as annual net revenues divided by total annual debt service (principal and interest). The result is expressed as a percentage of 100% and greater or as a factor of 1.0 and greater for general obligation and revenue bonds, respectively. Revenue bonds typically do not require voter approval, and the issuer typically has some flexibility in establishing a repayment schedule that meets the cash flow characteristics and needs of the utility. However, revenue bonds may have higher interest rates than general obligation bonds and may have marketing difficulties, particularly if the utility does not have an established earnings potential or a history of financial operations (AWWA, 2008).

5.2.3 Governmental Loans

Government loans from federal and state agencies typically are available to finance wastewater collection system capital projects. The Clean Water Act of 1987 established a State Revolving Fund (SRF) program to provide loans to wastewater utilities for capital improvements. These loan programs provide capital project financing at interest rates that typically are competitive with market rates. However, SRF programs often are competitive and dependent upon availability of funds. In addition, the timing and prioritization of the project will affect the suitability of SRF loans (AWWA, 2008). Therefore, the specifics of each loan program should be assessed to determine the suitability for financing wastewater collection system capital projects.

There are other variations and derivatives of bonds and loans that may be available to finance wastewater collection system projects. It is recommended that the reader refer to the capital financing references and suggested readings found at the end of this chapter, as well as local municipal and utility associations that typically provide links to this type of information.

6.0 REVENUE REQUIREMENTS AND FEE SETTING

Generating sufficient revenues to fund wastewater utility programs to properly operate, maintain, rehabilitate and replace wastewater collection system assets is an important aspect of wastewater collection system management. The following provides a summary of industry guidelines and common practices for establishing wastewater utility revenue requirements, user charges, and service fees.

6.1 Financial Condition

A wastewater system relies on the proper identification of annual revenue requirements to assess its financial condition and set user charges and service fees. By properly managing its annual revenue requirements and following industry guidelines for fee setting, a utility can maintain its financial condition. A financial management process can provide local officials with a high level of confidence that resources are being used properly. Financial management reports can provide local officials with the information they need to evaluate utility performance and set a course for future action. Through sound accounting procedures and periodic audits of financial records and procedures, assurance is gained that the resources of the utility are properly safeguarded. The following section presents recommended procedures for determining annual revenue requirements and setting defensible fees.

The financial management cycle for utilities is a process (see Figure 8.2) that has two major functions:

(1) Financial analysis, measurement, and planning provide feedback into the knowledge base of the service cycle to give managers and decision makers an understanding of the financial consequences of their decisions, both historically and prospectively. This information also is used to provide representations to third parties for accountability and financing purposes.

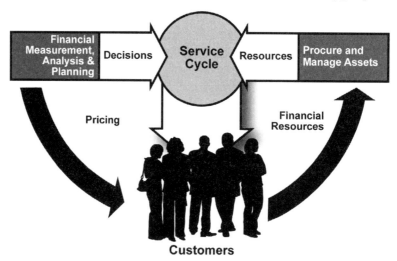

FIGURE 8.2 The financial management cycle. (Courtesy of Malcolm Pirnie, Inc. All rights reserved.)

(2) The process of financial analysis, measurement, and planning produces information that allows management and governing bodies to price the utility's services to recover costs (and for investor-owned systems, a fair return) in a fair and equitable manner. This allows the utility to bill for its services, collect revenue, and use that revenue to acquire, maintain, and manage its assets. The real asset in this exchange is the customer base, which provides demand for level of service and revenue to pay for it.

6.2 Establishing the Basis for Fees and Charges

To maintain levels of service, a wastewater utility must establish the revenues it needs and the fees and charges necessary to generate it. Two approaches typically are accepted for projecting total revenue requirements: the cash-needs approach and the utility approach. A third approach is the utility approach with cash residual. Each has a proper place in utility practice, and each, when properly handled, can provide for sound utility rate (or fee) setting. A fully detailed discussion is provided *Financing and Charges for Wastewater Systems* (WEF, 2005). The following summarizes the approaches and identifies the cost components for each.

6.2.1 Cash-Needs Approach

Under the cash-needs approach, a wastewater utility's revenue requirements equal its total cash requirements for a typical year of normal operations, or test year. The cash-needs revenue requirement is the sum total of the utility's O & M expenses, debt service costs (principal and interest payments), payments in lieu of taxes, other applicable taxes, and any capital outlays of the utility that are not funded with debt, capital reserves, or other outside sources. Revenues, other than those derived directly from user charges, are subtracted from the revenue requirement to determine the user charge. This final figure is the portion of the total revenue requirements that the utility must recover through its user charges.

Because of its focus on actual cash expenditures during the test year, the cash-needs approach tends to work well with government budgets, making it the most common method of calculating revenue requirements among government-owned wastewater utilities.

To summarize, basic revenue requirement components of the cash-needs approach include:

- Operations and maintenance expenses,
- Debt-service payments,

- Contributions to specified reserves,
- Cash-financed capital improvements, and
- Payments in lieu of taxes.

Depreciation is not included in the cash-needs approach. Expenses for O & M are projected based on actual expenditures and adjusted to reflect anticipated changes. Typically, O & M expenses include salaries and wages, fringe benefits, purchased power, other purchased services, rent, chemicals, other materials and supplies, small equipment that does not extend the useful life of major facilities, and general overhead.

The debt-service component of the cash-needs approach consists of principal and interest payments on bonds and loans. Capital expenditures are classified into three broad categories: normal accrual (routine) replacement of existing facilities; normal annual extensions and improvements; and major capital replacements and improvements. A utility should periodically review and update its needs in each of these areas to recognize changing conditions.

6.2.2 Utility Approach

The utility approach differs from the cash-needs approach in the way in which it considers capital costs. Instead of focusing on the cash outflows required to fund capital needs, the utility approach calculates capital costs using annual depreciation and a financial return that reflects the cost of capital on the adjusted used and useful utility plant in service, or rate base. To calculate the return component, the utility's cost of capital is multiplied by its rate base, which consists of the utility's used and useful net fixed assets and allowances for working capital and other adjustments (e.g., contributions in aid of construction).

The financial return component makes the utility approach the favored method of calculating revenue requirements for investor-owned utilities (and for those who regulate them). However, many government-owned utilities use the utility approach to calculate revenue requirements for extraterritorial customers. In these cases, a government utility may opt to use the cash-needs approach to calculate revenue requirements for customers inside its jurisdictional boundary and the utility approach for extraterritorial customers.

Utility-based revenue requirements may consist of:

- Operations and maintenance expenses,
- Depreciation expense,

- Return on rate base, and
- Payments in lieu of taxes.

For a government-owned wastewater utility (such as a sewer district or conservancy district), the total level of annual revenue required may be the same under either the cash-needs approach or the utility approach. The O & M expense component of total revenue requirements typically is the same under both approaches. Under the utility approach, the annual requirement for capital-related costs consists of two components: depreciation expense and return on rate base.

Each state commission and regulatory body has its own rules, regulations, and policies for determining total revenue requirements. Therefore, when preparing for any rate matter, it is essential that the utility follow procedures and policies of the regulatory body in establishing its revenue requirements.

As discussed above, revenue requirements are the total operating and capital costs of the system typically identified for a single year. User charge revenue requirements are net of any revenues the system receives from other sources; those costs that the utility must recover from its user charges and service fees to balance its cash sources and uses for the test year.

6.2.3 Selecting a Test Year

A test year can be defined as the annualized period for which costs are to be analyzed and user charges established. According to the WEF (2005), the test year for government-owned utilities can include a historical year, a historical year adjusted for known and measurable changes, or a projected year. Use of a historical test year is appropriate when a utility is in a period of normal and sustainable operations without unusually high capital costs related to infrastructure expansion or replacement. For growing utility, use of a projected test year becomes more appropriate because revenue requirements will more accurately capture the anticipated increases associated with capital and operating costs.

6.2.4 User Charge Revenue Requirements

The portion of annual system revenue requirements to be recovered through user charges depends on a utility's financing policy and other sources of income. To determine the amount of revenue that user charges must generate annually, the total revenue requirements must be reduced by nonrate or other system revenues. Other system revenues are defined as all revenues except those derived from wastewater user charges. Some examples of nonrate revenues include:

- Service fees;
- Penalties;
- Connection fees, tap fees, or system development charges; and
- Interest earnings.

Once the user charge revenue requirements of the system are determined, the costs can be allocated to system users. The costs to be allocated are referred to as costs of service as summarized in more detail by WEF (2005). Sound financial practices begin with determining the costs of serving the utility's customer classes. Communicating these costs to customers and designing user charges and service fees for recovery are the next steps to responsible financial management.

6.3 User Charge and Service Fee Types

Once rate revenue requirements are determined, wastewater user charges and service fees are established by utilities so that sufficient revenue is received from customers to ensure proper O & M, development, and perpetuation of the system and to maintain the ability to borrow. Rates, fees, and charges can be grouped into three broad categories:

- User charges or rates,
- Impact fees, and
- Other service fees.

6.3.1 User Charges

A wastewater utility's primary source of revenues is a user charge. User charges are the charges for service determined in a cost-of-service analysis. In a cost-of-service analysis, the cost responsibility of each utility customer class is determined through a cost-allocation process. Ideally, user charges will generate sufficient revenues to operate the system, preserve and build infrastructure, maintain financial integrity, and recover costs from customers equitably. Renewal and replacement of system assets is a critical component of revenues that need to be generated from user charges.

User charges take many forms depending on the community's goals and objectives. The basic components of user charges are fixed and volumetric charges. Fixed charges are charges per billing period that do not change with volume of wastewater flow. These charges often recover the costs of billing, general and administrative costs, a portion of debt service, and the costs of maintaining services.

Volumetric charges are those user charges that are based on volume of flow. These charges recover the remaining costs of service and are not limited to costs that vary according to flow. Volumetric charges can be uniform for all flows from all users or uniform by type or class of user (such as residential, commercial, restaurant, and hospital). Volumetric charges also can be tiered. Increasing volumes of flow can be charged at decreasing costs per unit (declining block rates) or at higher costs per unit (inclining block rates).

Finally, wastewater utilities that do not have volumetric charges which vary by customer type or class may assess "extra-strength" surcharges for industrial users. In this case, a customer's flow is monitored and compared to standard domestic strength wastewater. Surcharges typically are billed for each pound of constituent over a specified limit.

For a more detailed discussion on selecting user fee structures and developing rate schedules, see *Financing and Charges for Wastewater Systems* (WEF, 2005).

6.3.2 Impact Fees

Another significant funding source for wastewater utilities is impact fees. Impact fees, connection fees, system development charges, tap fees, plant investment fees, are all terms used to describe any fee or charge to recover capital improvement costs associated with system growth. These one-time charges are designed to recover the capital costs of growth from those causing the growth, rather than from the utility's existing customer base.

Without recovering investments in capital improvements for facilities expansion, the utility effectively would be subsidizing growth at the expense of existing ratepayers. For this reason, both existing and proposed investments in capacity are examined in calculating impact fees. The rationale for such fees is the unrecovered investment in available capacity, whether that capacity exists or is proposed.

In charging new customers for both past and new investments in capacity, the impact fee, like other such fees, promotes a concept in utility ratemaking called "intergenerational equity". Intergenerational equity means that existing customers do not subsidize new customers and vice versa. In many communities, this is often referred to as "growth pays for growth". Impact fees can be designed to minimize the subsidization of new growth.

Three approaches for calculating impact fees are commonly used for wastewater systems: buy-in approach, incremental cost approach, and combined approach (WEF, 2005). Each approach results in impact fees that are used to pay for growth-related facilities such as treatment plants, collection and interceptor mains, and lift stations. Revenues from these fees should be incorporated into the utility's financial management strategy.

6.3.3 Other Fees

To balance its sources and uses of funds, utilities often assess miscellaneous fees. Such fees range include basic administrative charges for service, turn-on/turn-off fees, penalties for delinquent payments, and miscellaneous service charges, such as standby charges, availability of service fees, and installation or hookup fees. If a wastewater collection system has a reuse system component in its operations, charges for reuse or reclaimed water also may be appropriate. As in determining the user fees, establishing miscellaneous fees should be based on the costs of providing the service.

7.0 ROLE OF GOVERNMENT ACCOUNTING STANDARDS BOARD STATEMENT NUMBER 34 IN BUDGETING AND FINANCIAL PLANNING

As wastewater utilities establish user charges and service fees to support capital and O & M programs, they also will need to document financial results—an important aspect of the budgeting and financial planning process. This section reviews requirements of GASB 34. It discusses how budget priorities and asset-management programs are represented in financial reporting requirements for wastewater utilities.

The GASB is a private, nonprofit organization formed in 1984 to develop and improve accounting and reporting standards for state and local governments (www.gasb.org). Between 2001 and 2006, the GASB phased in the requirements of GASB 34, which was aimed at providing more comprehensive financial reporting information about the ability of governments to repay ongoing, increasing costs of infrastructure replacement debt and to continue to provide service. A discussion of GASB 34 requirements is pertinent to wastewater collection system management because it affects wastewater utility financial reporting requirements and provides an optional financial reporting approach that is based on asset-management practices described in other chapters of this manual.

The GASB 34 requires that all government agencies use one of two possible accounting approaches: (1) the conventional approach or (2) the modified approach. Many GASB 34 requirements are for general governments, such as cities and counties, which rely on general funds to support their asset-management activities. Before GASB 34, these governments did not list and depreciate infrastructure assets on their financial statements. Most wastewater agencies and governments with enterprise funds already listed assets on their balance sheets; so the conventional approach under GASB 34 imposed only minimal additional requirements on the utility.

Under the GASB 34 conventional approach, wastewater utilities need to satisfy the following requirements (www.gasb.org):

- Ensure that all infrastructure assets are included on the balance sheet and are depreciated using the accrual method of accounting.
- Include management's discussion and analysis as a narrative to the financial statements, which discloses significant financial activities, events, and conditions that affect financial position.
- Compare actual yearly expenditures to original budgets, not revised budgets.

These requirements are important to understand as capital and O & M programs are being developed and budgets are prepared. Necessary information can be gathered throughout the year to demonstrate results that programs and budgets were intended to achieve.

The modified approach to satisfying GASB 34 permits the current cost of maintaining assets to be reported on the financial statements in lieu of depreciation expense under several conditions. Under the modified approach wastewater utilities need to satisfy the following requirements (www.gasb.org):

- Maintain an up-to-date inventory of eligible infrastructure assets,
- Establish a target condition level for the assets,
- Complete a condition assessment at least every three years,
- Implement a formal asset-management system,
- Estimate and budget necessary funding to maintain assets at the target condition level,
- Conduct a review of the effectiveness of maintenance procedures,

- Ensure that assets are being maintained at or above the target condition levels, and

- Compare actual preservation expenses to budgeted preservation expenses.

If a wastewater utility complies with all of the requirements of the modified approach, it can report assets on a preservation basis, and may be able to reduce the amount of depreciation expense reported on its financial statements. This is important because depreciation expense represents a large, noncash cost to a utility that (1) reduces net income and (2) reduces the asset value on the balance sheet by the amount of the accumulated depreciation calculated because the asset was placed in service for utilities that use the accrual basis of accounting. By reducing depreciation expense, a utility will maintain a higher value of assets on their balance sheets, thus increasing net equity, and reducing the debt-to-equity ratio and reported leverage. As a result, the management practices that must be followed to use the modified approach could improve the reported financial condition of the utility. It also could improve the rating agency's view of the utility, although ratings are based on several factors.

Similarly, wastewater utilities that use the accrual basis of accounting—and to the extent that it has been expending funds to keep assets in good condition so that they are depreciating more slowly than measures of accounting depreciation would predict—the modified approach allows the utility to report the cost of maintaining the assets as an expense, rather than as an accrual of depreciation expense. In the short term, the modified approach could reduce utility expense and increase net income. Perhaps the most important benefit of the modified approach is that it requires a utility to implement "best practices" in the management of its infrastructure assets.

The additional cost required to implement the modified approach, however, is not insubstantial. For example, completing a condition assessment every three years and on assets tracking and updating condition would be costly but necessary under this approach. Regardless of which approach is used, however, management of assets on a proactive and preservation basis will provide long-term financial benefits. In addition, documentation of progress and results of capital and O & M programs is an important part of the asset-management process.

8.0 SELLING THE ASSET-MANAGEMENT APPROACH AND BUDGET PLAN

Utility management now has basic information to develop a long-term financial management plan. The key here is to "think long-term, act short-term". By putting the cost effects of the asset-management program into the context of comprehensive financial dynamics, a clear understanding of the future financial consequences of current management decisions emerges.

These steps create help inform decision making:

- Prepare a long-term forecast. Most utilities now use corporate planning models to develop financial plans. These range from simple spreadsheets to complicated commercial software packages. However, too often these are not incorporated into the budget process, which focuses on immediate needs and effect. It is essential that long-term forecasts be incorporated into the budget process to identify future "rate spikes" and plan to mitigate them over a longer period of time. This can include efficiency improvements, inflationary rate indexing to generate sufficient funds slowly over time, and changing capital improvements scheduling.

- Clearly demonstrate that costs drive revenue requirements. The forecast should identify the cost factors that create the need for additional revenue. In general, there are five major cost categories that should be displayed: operating costs (including taxes); cash flow requirements; reserve requirements; capital improvement/debt service; and debt service coverage.

- Develop and demonstrate several scenarios. The financial planning process requires an understanding of the range of outcomes resulting from various economic conditions. Providing a worst case, optimal, and medium scenario can often facilitate the decision-making process. A range of assumptions regarding cost inflation, interest rates, financing terms, customer growth, asset failure rates, and other relevant factors should be explored. However, the factors adjusted within each scenario should be limited to keep comparisons clear and easy to compare and understand. Office productivity software, such as spreadsheets, can illustrate these factors easily and clearly. Graphics can help display the results in way that is easy to understand.

- Work interactively. A summary results page of the planning model can be projected onscreen at budget meetings, allowing scenarios to be explored interactively and alternative assumptions tested on the spot. This provides also provides a way for all parties to contribute to the effectiveness of the process.

- Keep the discussion in context. Whenever short-term solutions are suggested, both short-term and long-term effects on user charges should be shown.

- Be ready with consequences. The big question is always, "So what?" Seldom expressed so succinctly, it is nevertheless on everyone's mind and will be expressed in indirect ways. The development of consequences documentation is essential to providing a basis of confidence for the final budget decision.

Asset management provides an effective framework for setting budget priorities and addressing the challenges of infrastructure renewal and replacement. It is equally important that these determinations be integrated effectively into the budget process. By using an interactive approach incorporating long-term forecasting with short-term budgeting, a productive forum is created that helps to balance the needs of deteriorating infrastructure with rate effects. An understanding of the necessity of that balance can only be provided by a long-term perspective.

9.0 REFERENCES

American Water Works Association (2008) *Water Utility Capital Financing*, 3rd ed.; Manual of Practice No. 29; American Water Works Association: Denver, Colorado.

Brealey, R. A.; Myers, S. C. (2000) *Principles of Corporate Finance*, 6th ed.; McGraw-Hill: New York.

Bureau of Labor and Statistics. Consumer Price Index. http://www.bls.gov (accessed October 2008).

Department of the Interior, Bureau of Reclamation (2007) *Change in Discount Rate for Water Resources Planning. Fed. Regist.*, **72** (221), Nov 16.

Government Accounting Standards Board Home Page. http://www.gasb.org (accessed December 2007).

Water Environment Federation (2005) *Financing and Charges for Wastewater Systems*, Manual of Practice No. 27; McGraw-Hill: New York.

Chapter 9

Safety, Standard Procedures, Training, and Certifications

1.0	INTRODUCTION TO AND APPLICABILITY OF THE OCCUPATIONAL SAFETY AND HEALTH ADMINISTRATION 203		3.5	Standard Operating Procedures 210
			3.6	Employee Orientation and Training 210
			3.7	Workplace Inspections 212
2.0	APPLICABILITY OF HEALTH AND SAFETY REGULATIONS 204		3.8	Emergency Medical and First-Aid Procedures 213
			3.9	Medical Aid and First Aid 213
3.0	HEALTH AND SAFETY PROGRAM ELEMENTS 204		3.10	Health and Safety Program Promotion 214
	3.1 Health and Safety Policy 205		4.0	SAFETY PROGRAM ORGANIZATION 214
	3.2 Individual Responsibility 207		5.0	SAFETY EQUIPMENT 215
	3.3 Safety Committee 207		6.0	SUMMARY 216
	3.4 Health and Safety Rules and Written Programs 208		7.0	REFERENCE 216

1.0 INTRODUCTION TO AND APPLICABILITY OF THE OCCUPATIONAL HEALTH AND SAFETY ADMINISTRATION

The Occupational Safety and Health Act of 1970 was enacted by Congress to ensure safe and healthful conditions for working men and women by authorizing enforcement of the standards developed under the Act; assisting and encouraging the States in their efforts to assure safe and healthful working conditions; and providing research, information, education, and training in the field of occupational safety and health.

The Occupational Health and Safety Administration (OSHA) was assembled to carry out the Act. Many states have OSHA-approved state plans and, therefore, may have more stringent regulatory requirements that must minimally meet federal OSHA standards. Each employer is required to provide a safe workplace and job that is free from recognized hazards that are causing, or are likely to cause, death or serious physical harm under the General Duty Clause.

The OSHA regulations are contained in the Code of Federal Regulations (CFR), volume 29. Part 1903, Inspections, Citations and Proposed Penalties, and part 1904, Recording and Reporting Occupational Injury and Illnesses, apply to all employers with some exception for employers with 10 or fewer employees in a calendar year. There are two other parts within 29 CFR that are applicable to the wastewater collection systems industry: General Industry Standards, part 1910, and Construction Standards, part 1926. Any time a construction-based activity is being conducted, the construction standard will apply. Refer to OSHA applicability sections for further explanation.

2.0 APPLICABILITY OF HEALTH AND SAFETY REGULATIONS

Determining which regulatory requirements apply can be a daunting task and should be carried out by a designated safety professional that has had the appropriate training. Both the General Industry and Construction standards will apply to many different operations in wastewater collection systems. Each part of the regulation begins with an applicability section that will help the safety professional determine if the section applies. The Occupational Health and Safety Administration has programs such as consultative services that also may be used as a resource for determining applicability.

3.0 HEALTH AND SAFETY PROGRAM ELEMENTS

A health and safety program is a defined plan of action designed to protect employees in the workplace. The OSHA legislation requires that all workplaces implement some form of a health and safety program that meets or exceeds the minimum elements set forth in regulations. Minimum program elements may include (OSHA, 2007):

- Health and safety policy,
- Individual responsibility,
- Health and safety committee,
- Health and safety rules through written programs,
- Standard operating procedures (see Figure 9.1),
- Employee orientation and training,
- Workplace inspections,
- Emergency procedures,
- Medical aid and first aid,
- Health and safety promotion, and
- Workplace specific items.

Because each utility may differ in size, type of operations, and resources, one program cannot be developed for all wastewater collection systems. It is important that each utility develop a health and safety program that is specific to its needs. This chapter discusses general health and safety guidelines but should not be considered all inclusive.

3.1 Health and Safety Policy

In addition to a program, every organization should have a health and safety policy. The policy should be a statement of philosophy and guidelines to which the organization and senior management are committed. The health and safety policy should receive the same level of enforcement and consideration as all other policies. Senior management is responsible for ensuring that the health and safety policy is carried out and enforced.

A health and safety policy does not have to be lengthy; in fact, it is better written clearly and concisely. The policy should contain a statement of management's commitment to the safety and health of employees, the organization's philosophy, and the general objectives of the program. It should also identify the person responsible for the program and state that all employees are responsible for adhering to safety rules and regulations. It is important to include a statement that poor safety performance will not be tolerated. The policy should be signed by the director of operations, kept up to date, and communicated to all employees.

STANDARD OPERATING PROCEDURES

Your Company Name Here	MANUAL			
SUBJECT Name of SOP	NUMBER	REV	EFFECTIVE DATE 6/2005	PAGE OF
	SUPERSEDES	PREPARED BY		APPROVED BY

1.0   Purpose:

2.0   Organization affected:

3.0   Reference:

4.0   Hazard description:

5.0   Preventive measures:

6.0   Engineering controls:

7.0   Protective equipment:

8.0   Personal hygiene and sanitation:

9.0   Hazard communication and training:

FIGURE 9.1   Sample standard operating procedure.

3.2 Individual Responsibility

Individual responsibility for health and safety applies to everyone in the workplace, including top-level management. Each person is obligated to abide by established rules and regulations as they perform their job duties, which is why everyone must understand their responsibilities. Including specific health- and safety-related responsibilities in job descriptions can ensure that an individual understands what is expected.

Responsibilities for workers, first-line supervisors, management, safety coordinators, and the safety committee should be identified. Authority for enforcement and the level of enforcement also should be identified. This can be accomplished by developing an organizational flow chart for the program. Although responsibility and authority can be delegated to subordinates, management has ultimate accountability to ensure that the program is carried out.

3.3 Safety Committee

Establishing a safety committee is an integral step to creating an effective safety program. A safety committee encourages employee and management involvement and is a great tool for increasing awareness and imparting a sense of ownership. The safety committee should have an organized structure, a clear statement of purpose (committee charter), and established procedures for conducting regular meetings. Committee members can be appointed on a rotating schedule or can be made up of volunteers. Members should include management and all levels of workers. The safety committee also should be empowered through a clearly defined level of authority and resources to affect change. Some examples of duties that a safety committee may carry out are:

- Reviewing and investigating accident report, determining responsibility, and making recommendations for prevention;
- Reviewing statistics on lost time and no-lost time accidents, near-misses, first aid, and preventable and non-preventable vehicle accidents;
- Periodically reviewing completed tailgate talks, safety meetings, and worksite inspections;
- Developing and reviewing job hazard assessments (JHAs) (see Figure 9.2);
- Developing a training program;
- Reviewing written safety programs; and
- Reviewing operational changes for safety issues.

Job title:	Job location:	Analyst:	Date:
Task #	Task description:		
Hazard type:	Hazard description:		
Consequence:	Hazard controls:		
Rationale or comments:			

FIGURE 9.2 Sample job hazard analysis form.

Attention to trends in the above statistics can be useful in promoting safety awareness at all levels. Essentially, the committee should be active in the development, implementation, and monitoring of all phases of the health and safety program.

3.4 Health and Safety Rules and Written Programs

As mentioned earlier, minimum health, and safety regulations have been established by OSHA. States with OSHA-approved state plans may include regulations that are more stringent than the federal standard. Using these established regulatory requirements as minimum guidelines, a workplace must develop a written health and safety program. The workplace may create a written program that is at least as stringent as both the federal and state regulations. The written program must be specific to the industry and workplace and should include safe standard operating procedures (SOP) to achieve a safe and healthful workplace. Standard operating procedure development will be discussed later in this chapter.

Each organization should have developed written safety rules that are readily available and communicated to all employees. Many resources are available for assistance in developing these written programs. A list is included at the end of this chapter. Basic programs that typically are required in the wastewater collection system industry are listed below.

- Overall written safety program that establishes safety policy, organizational authority, and disciplinary action;
- Excavation and trenching;
- Work zone safety;
- Permit-required confined space;
- Heavy equipment operation and rigging;
- Hazard communication;
- Electrical safety lockout/tagout;
- Underground utility safety;
- Asbestos pipe;
- Personal protective equipment (PPE);
- Walking working surfaces/fall protection;
- Workplace exposure to wastewater;
- Respiratory protection program;
- Hearing conservation program;
- General safety awareness (nature exposures, heat/cold stress, etc.);
- Emergency action plan; and
- Machine guarding.

3.5 Standard Operating Procedures

An SOP should be created for each critical task using a process called a JHA, sometimes also referred to as job safety assessment (JSA). A JHA is the first step in creating an SOP. A job is selected and then broken down into steps. Each step of the task is evaluated to identify existing and potential hazards encountered or developed during the task. After risks have been identified, preventative measures are defined. The person who carries out the task should participate in the JSA process because he or she is most familiar with the job and potential safety risks and can help develop realistic preventative or corrective measures.

Standard operating procedures are an important tool for all employees because they establish and ensure consistency and safety throughout a task and provide a for quality assurance. An SOP should be used to train new employees and as a reference for infrequently performed tasks. An example of an SOP is included in Figure 9.1. Some typical SOPs that should be developed in the wastewater collections system industry are listed below.

- Permit required for confined-space entry;
- Inspection of high risk mains;
- Lockout/tagout;
- Roding and jetting of mains;
- Electrical installations and safety;
- Work zone and traffic control;
- Easement maintenance and debris removal;
- Operation of heavy equipment; and
- Excavation and trenching (specific to a task; for example, installation of trench box).

3.6 Employee Orientation and Training

Health and safety education is an integral part of new employee orientation. Statistics show that inexperienced workers are likely to be involved in accidents more frequently than experienced staff. Generally, orientation will cover general policies, administrative items, and organizational relationships. Employee orientation should also include a distinct module related to safety such as emergency procedures, location of first-aid stations, location of health and safety rules and regulations, employee responsibilities, and expectations in safety, reporting accidents, near misses, injuries and unsafe conditions. It also should cover use of PPE and the right to refuse to perform hazardous work if unsafe conditions exist.

General new employee orientation should not replace job and site-specific training. A new employee can be expected to absorb only a certain amount in the first few days, so it is beneficial to set up a "mentor" that the employee can go to with any questions. The department supervisor or safety coordinator should establish an in-depth safety orientation as part of job-specific training. This can include developing a safety training awareness schedule, reviewing applicable JHAs, SOPs, and PPE requirements, and establishing guidelines for on-the-job training. This level of orientation does not take the place of more in-depth training courses. The new employee should attend such courses as soon as possible.

In-depth safety training should follow orientation. An employee's failure to recognize hazards is one of the most frequent causes for serious injuries and illness in the industry. Hazard-specific training is required by OSHA as identified in the regulations and in written programs. Regulatory requirements mandate the type and frequency of training, which should be documented in the written program. Documentation of the written program, the training program, and training records are necessary to achieve regulatory compliance. The written programs and SOPs listed in this chapter are typical areas of concern in wastewater collections systems that require training (see Health and Safety Rules and Written Programs and Standard Operating Procedures sections). The lists are not meant to be all inclusive of training requirements; a safety professional should be consulted for further guidance.

The objective of a safety training program is to effectively raise awareness and skill levels to an acceptable level while easing the incorporation of policies and procedures into a specific job function. Different levels of training should be established based on job responsibility for managers, supervisors, trainers, and the worker. Developing a training matrix is an excellent way to track training requirements by position. Although training frequency varies, it should occur when there is a new employee, a transfer to a new job, changes to equipment or processes, or when inadequacies are demonstrated. Training frequency should be identified in the written program.

Training can be accomplished through several avenues such as weekly "tailgate" meetings, regular monthly meetings, and specialized sessions. Weekly "tailgate" meetings are typically an informal training session held within a work group that covers a wide spectrum of safety topics and usually lasts approximately 10 minutes.

Monthly training sessions may be delivered department wide and typically cover generalized topics, lasting approximately 20 to 30 minutes. Staff members who are knowledgeable on the subject, such as a supervisor or safety coordinator, may conduct the monthly sessions. Specialized training is generally more detailed and specific, can last from one hour to one week or more, and is instructed in a more formal setting by a qualified instructor. If the facility has an onsite staff safety officer, training coordinator, or other qualified individual, that individual can deliver the appropriate training. Training also can be conducted on- or off-site by an outside qualified vendor such as the National Safety Council or other industry organization. Written training modules should be developed for each topic being covered (other than for tailgate-style meetings). Modules include goals and objectives of the course, and items to be covered including applicable regulations, summary, and test.

A variety of training techniques should be incorporated into the session to encourage employee involvement and address different styles of learning. One example would be to use video clips and flip charts for discussion groups. It may also be required to have a practical exercise such as with trenching and excavations or confined space entry. In this case, the instructor would arrange to carry out field exercises in which employees could participate in mock activities such as soil classification or entering a confined space. The employee should be able to demonstrate correct performance and knowledge of the topic by the end of the training session.

3.7 Workplace Inspections

Another key aspect in a safety program is regular, unannounced workplace safety inspections. Periodic inspections are a regulatory requirement that will help to identify and correct any existing or potential hazards or unsafe behaviors. For periodic inspections to be successful, targets should be set and monitored. Directors, managers, supervisors, and team leaders should be assigned a minimum number of safety inspections to complete throughout the year. Any variance from the target should be discussed at the each individual's year-end performance evaluation. The workplace includes the main and remote operations locations and mobile job sites. Some items to include in inspections are work conditions and procedures; equipment, machinery, and tools; personal protective equipment; electrical safety; housekeeping; and adherence to SOPs and regulatory requirements.

Many other items may be included in the inspection and should be specific to the area or job being inspected. Any time that a danger to life or health is encountered, corrective action should be taken immediately. All inspection items should be documented, deficiencies noted, and corrective actions implemented. Inspections can be performed by a team or a designated individual that is knowledgeable of regulations and procedures (such as the safety committee or supervisor).

3.8 Emergency Medical and First-Aid Procedures

Emergency procedures are plans specific to the workplace for incidents such as fire, explosion, flooding, severe weather, chemical releases, or other events that may endanger personnel and require immediate response and decision making. The objective of the plan is to protect life and human health, protect critical infrastructure, and allow quick return to normal operations. Communication, training, and periodic drills are required to ensure effective implementation of the plan. Emergency preparedness and procedures are discussed further in Chapter 10.

3.9 Medical Aid and First Aid

Medical aid and first aid can apply in several different ways in the wastewater collection systems industry. First, as part of before starting work, employees are generally required to establish a health "baseline" by completing a general physical. More extensive medical testing may be required for certain job tasks such as the wearing of a respirator. If there are job exposures that are at or above established OSHA guidelines, employees must be enrolled in a medical monitoring program. A medical monitoring program may include annual hearing tests as part of the hearing conservation program, annual pulmonary function testing for respirator wearers, or annual blood work for possible chemical exposures. The safety coordinator or other designated person should work with the occupational health nurse to establish a medical monitoring program as applicable.

Second, procedures and policies for medical services should be established for employees involved in an accident or who have been injured. The policy should include procedures for reporting, obtaining medical care, and returning to work including provisions for light duty or modified duty. The OSHA standards require that medical services be available "in near proximity," which has been defined as three to four minutes. If an organization does not have access to medical services within these provisions, they must be provided. If an organization has designated first-aid providers, the written program should identify those responders, the location of first-aid stations, and guidelines for type and frequency of training.

Immunization programs, as recommended by the occupational health nurse or as required by OSHA, for service operators should be followed. Guidelines for employees exposed to wastewater in the eyes, mouth, ears, or through complete immersion should be accessible and well known to all employees. Health education must be provided to all employees. Shower facilities, special soaps, nail brushes, and other personal hygiene supplies also should be provided. The blood borne pathogens standard covers wastewater workers if they are exposed to raw blood or other sources of infectious materials from hospitals or mortuaries. Employees exposed to sharps such as needles also are covered under the standard. Simple approaches, such as baseline health assessments before employment and uniform service with cleaning, are becoming standard practices in the wastewater industry.

3.10 Health and Safety Program Promotion

Much controversy exists regarding how health and safety programs are promoted. Many organizations have established safety incentive programs or recognition programs that award employees for good safety practices or for those who remain accident free. This causes concern among some who believe that this type of award or incentive program promotes under reporting of accidents and encourages the "walking wounded." Positive promotion without encouraging this type of activity can be accomplished through other tactics. For example, establishing a point system for those who attend the most safety meetings, include recommendations in near-miss reports, generate innovative safety ideas, or participate on safety committees is just one method that encourages proactive safety behavior and not under reporting. Promotion of the health and safety program should be well thought out and should include positive message that is fair and equal.

Health and safety minimum program elements are regulated and established by OSHA although a state or organization may choose even more stringent rules. The elements discussed above are based on federal OSHA and are not intended to be all inclusive. More specific elements may be needed to address specific workplace activities. Utilities should consult with a safety professional or other agency such as OSHA's consultative services to ensure program compliance.

4.0 SAFETY PROGRAM ORGANIZATION

The safety program should be considered a key business element and have full support and funding from management. Program success ultimately is the responsibility of upper management. If the system is large enough, a safety coordinator should be appointed who directly reports to upper management and works with the department managers, front-line supervisors, and employees to implement the program. The safety coordinator should have the authority to award employees or apply disciplinary and corrective action as needed with support from management. Often departments may not be large enough to have a fulltime safety coordinator. In such cases, the responsibility ultimately falls on the front-line supervisors. Any person who is assigned responsibility for the program should have the support of management, including the time, training, and authority for safety.

5.0 SAFETY EQUIPMENT

Availability of approved safety equipment is necessary to maintain safety and health of employees. In the wastewater collection system, the tasks that are performed and associated hazards will dictate the equipment needed. Personal protective equipment and mechanical protective equipment are two types of safety equipment regularly encountered.

All PPE must be approved and rated for hazards encountered. Certain PPE must also be fitted to the person wearing or using the PPE. There are two agencies that typically approve PPE: ANSI and the National Institute of Occupational Safety and Health (NIOSH). Safety glasses, steel-toe shoes, and hard hats are a few of the items that ANSI regulates. Other equipment includes eyewashes and showers, railings, and sound-level meters. The NIOSH generally sets standards for items such as respirator cartridges, dust-mask-style respirators and other occupational exposure criteria, such as the permissible exposure level. Mechanical protective equipment standards for items such as machine guards, valve protection, some ventilation systems, and man-lifts also are established by ANSI.

A comprehensive assessment of facilities and field activities should be conducted by a qualified individual to determine what type of PPE or mechanical protective equipment may be required. Generally, all types of protective equipment should be regularly inspected and maintained. The Occupational Safety and Health Administration has incorporated inspection requirements for equipment. If there is no regulatory guidance for inspections, maintenance, or replacement, the manufacturer's recommendations should be followed. A documented annual inventory and inspection is recommended at a minimum for all issued and stock PPE.

6.0 SUMMARY

Developing, implementing, and maintaining an effective health and safety program is a challenging task. The items discussed in this chapter are offered as guidelines as there is far too much information and specific applicability issues to provide a comprehensive outline. However, the sections addressed should provide a solid starting point to help in developing a thorough health and safety plan for a wastewater collection system. A strong safety program starts with management support and employee involvement.

7.0 REFERENCE

U.S. Department of Labor, Occupational Health and Safety Administration (2007) Safety and Health Management System eTool. http://www.osha.gov/SLTC/etools/safetyhealth/mod1.html (accessed October 2007).

Chapter 10

Emergency Preparedness and Security

1.0 INTRODUCTION AND OVERVIEW 218	3.4 Establish Relationships with Other Local and State Response Agencies 223
2.0 DISASTER AND SECURITY RISKS AND THEIR POTENTIAL EFFECTS ON WASTEWATER COLLECTION SYSTEMS 219	3.5 Develop an Understanding of the National Incident Management System and the Incident Command System 223
2.1 Natural Disasters 219	
2.2 Incidents Caused by Humans 219	3.6 Be Aware of Cyber-Vulnerabilities 224
2.3 Technological Failures 220	3.7 Enter into Mutual Aid Networks and Agreements 224
3.0 PREPAREDNESS PLANNING ACTIVITIES FOR WASTEWATER COLLECTION SYSTEMS 220	3.8 Develop Crisis Communications Plans 225
3.1 Develop an Organizational Culture of Emergency Preparedness and Security 221	3.9 Conduct Training and Exercises 225
	3.10 Stay Abreast of Developments in the Industry 226
3.2 Prepare Employees for Disaster Incidents 221	4.0 PHYSICAL PREPAREDNESS FOR WASTEWATER COLLECTION SYSTEMS 226
3.3 Develop Viable Emergency Response Plans 222	

(continued)

4.1	Provide for Emergency Electrical Power	227		Emergency Operations Centers	230
4.2	Ensure Adequate Protection of Mains from Waterways	227	5.2	Activate Crisis Communications Plans	230
4.3	Reinforce Vulnerable Mains	228	5.3	Check on the Welfare of Employees	231
4.4	Locate Pumping and Lift Stations above Flood Elevations	229	5.4	Conduct Damage Assessments	231
4.5	Protect Vehicles and Heavy Equipment	229	5.5	Communicate and Coordinate with Other Local and State Response Agencies	231
4.6	Provide for Necessary Repair Equipment and Materials	229	5.6	Activate Mutual Aid Networks and Agreements	232
4.7	Ensure the Availability of Reliable Communication Systems	229	5.7	Make Emergency Repairs	232
5.0	DISASTER RESPONSE IN WASTEWATER COLLECTION SYSTEMS	230	5.8	Maintain Records to Maximize Federal Emergency Management Agency Reimbursement	232
5.1	Activate Emergency Response Plans and Open		6.0	SUGGESTED READINGS	233

1.0 INTRODUCTION AND OVERVIEW

Incidents in recent years have highlighted the potential effects of natural incidents and damage caused by humans to utility infrastructure, including wastewater collection systems. Although much of the focus has been on water systems, treatment plants, and other infrastructure, wastewater collection systems also have proven to be vulnerable and their failure can result in significant consequences on public services and the environment. Regulatory changes also have increased the importance of preventing failures in wastewater collection systems. Wastewater collection system managers should put in place several measures to prepare for disaster incidents. Although some of these measures may apply more to larger wastewater collection systems, most of the guidance in this chapter is scalable to systems of all sizes.

2.0 DISASTER AND SECURITY RISKS AND THEIR POTENTIAL EFFECTS ON WASTEWATER COLLECTION SYSTEMS

Potential incidents that can affect wastewater collection systems, for which wastewater collection system managers and operators must be prepared, can be grouped into three categories: natural disasters, incidents caused by humans, and technological failures. Each of these three categories is discussed in greater detail below.

2.1 Natural Disasters

Natural disasters that can affect wastewater collection systems include hurricanes, tornadoes, high-wind incidents, floods, flash-floods, landslides, ice storms, extreme winter weather, earthquakes, pandemics, and others.

Flooding from natural disasters has the potential to sever or washout buried and aerial mains and other assets, to wash silt and other debris into manholes and mains, and to flood and damage pumping and lift stations. Winds, especially in conjunction with heavy rains, can cause mains and other assets to be damaged by uprooted trees. High winds and flooding have the potential to seriously damage and even destroy pumping and lift stations. In wind and flooding incidents, manholes and some other system components can be buried under silt and debris. Earthquakes pose a threat of breaks to mains and structural damage to other facilities.

All natural disasters have the potential to cause failures in the supply of electrical power to pumping and lift stations and communications systems and to disrupt supervisory control and data acquisitions (SCADA) systems.

All forms of natural disasters also have the potential to damage vehicles, heavy equipment, and other assets. Finally, many incidents can cause sanitary sewer overflows (SSOs) and combined sewer overflows, which can result in further damage to assets and to the environment.

2.2 Incidents Caused by Humans

Incidents caused by humans can be the result of unintentional or intentional acts or vandalism by disgruntled employees or other individuals.

Unintentional effects by humans on wastewater collection systems can include construction excavation and drilling damage to mains, accidental chemical spills into manholes and mains, vehicular accidents involving pumping and lift stations or other above-ground facilities, and employee errors.

Among the possible motives of individuals attempting to intentionally damage wastewater collection systems are to get attention among their peers, extract revenge for perceived wrongdoing, create an opportunity to be a hero, cause environmental damage, impair water supplies downstream, cause economic disruption, or attack a specific location through the wastewater collection system. Intentional damage caused by humans to wastewater collection systems can include main blockages and subsequent SSOs; damage to components, including mains, manholes, and pumping and lift stations; cyber-hacking or cyber-terrorism to shut down pumping and lift stations; contamination of the collection system; or use of some part of the system, such as chlorine, to cause injuries or damage.

2.3 Technological Failures

Both natural disasters and incidents caused by humans result in the technological failures listed above. Technological system components also are subject to intrinsic failures and loss of electrical power supplies, which can result failures of pumping and lift stations, SCADA systems, and communications systems or breaks and structural failures of mains and other facilities.

All of the above types of incidents require that wastewater collection system managers conduct the planning and physical preparedness measures as discussed in the following sections.

3.0 PREPAREDNESS PLANNING ACTIVITIES FOR WASTEWATER COLLECTION SYSTEMS

Wastewater managers should implement several preparedness planning measures in anticipation of the effects discussed above:

- Develop an organizational culture of emergency preparedness and security.
- Prepare employees for disaster incidents.
- Conduct a vulnerability assessment to potential identify hazards.
- Develop viable emergency response plans (ERPs).

- Establish relationships with other local and state response agencies.
- Develop an understanding of the National Incident Management System (NIMS) and the Incident Command System (ICS).
- Be aware of cyber-vulnerabilities.
- Enter into mutual aid networks and agreements.
- Develop crisis communications plans.
- Conduct training and exercises.
- Stay abreast of developments in the industry.

3.1 Develop an Organizational Culture of Emergency Preparedness and Security

The first step is to develop and promote awareness of what can happen in a system during and after a disaster. The preparedness process should include worst-case scenario. As Hurricane Katrina demonstrated in New Orleans, Louisiana, in 2005, worst-case-scenario planning should be pushed to the most extreme situation imaginable. Utility system managers should be careful not to overlook wastewater collection systems in their preparedness process. Utility systems should foster a security/disaster awareness culture among employees and provide training and reinforcement in various forms and at all levels.

3.2 Prepare Employees for Disaster Incidents

Many utility systems have learned that they also must address employee concerns when responding to a security incident or other disaster. Addressing these issues can determine the success of an incident response and benefit long-term employee relations.

Employees are a wastewater collection system manager's most valuable, and sometimes most vulnerable, asset in a disaster. Helping them prepare for disasters includes:

- Educating them in the agency's plan;
- Assigning special disaster responsibilities, training, equipment and personal protective equipment;
- Apprising them of their role and what will be expected of them;
- Providing them with the necessary training, tools, and safety equipment;

- Establishing policies for mandatory work and criteria for leave;
- Providing shelter for employees with food, water, provisions for sleeping, and backup electrical power;
- Providing similar facilities and telephone hotlines for their families;
- Providing guidance for preparations at home; and
- Providing assistance in dealing with insurance and related issues associated with personal property damage.

Other measures to support employees include counseling as needed, paycheck cashing services, and post-disaster recognition and celebration, where applicable.

3.3 Develop Viable Emergency Response Plans

Water and wastewater utilities need to develop an ERP that incorporate all details necessary for effectiveness. Emergency operations plans and business continuity plans sometimes are used synonymously with ERPs.

The first step in developing an ERP is to identify the specific types of incidents that may affect a particular wastewater collection system. Emergency response planning requires detailed consideration of all potential disaster events. The plan should cover all employees and provide the level of detail necessary in a real event, including the availability of information on system infrastructure. Operational staff members need to be involved in assembling the details of the ERP to ensure their effectiveness.

Employees, representatives of other local agencies, and potential mutual aid providers are only as valuable as a collection system manager's ability to contact them during or following a disaster. Contact information can be the weak link in the disaster response chain. Managers must maintain this information in a non-electronic form that is current, accessible, and durable.

Emergency response plans should include:

- Necessary internal and external contact information and details on the system infrastructure where that information is accessible during and after disaster incidents;
- Provisions for materials and equipment that may be needed after a disaster;
- Maps and other critical system information; and
- Succession plans, in the event that key personnel are unavailable.

As Hurricane Katrina demonstrated, all local agencies must be prepared to provide for their own needs for at least three days following a disaster, before aid may be available from other agencies.

All personnel need to be trained in the details of their ERP. Emergency plans are not static documents and wastewater collection system managers must continue to monitor their vulnerabilities and refine and adjust their response plans.

3.4 Establish Relationships with Other Local and State Response Agencies

Wastewater collection system managers should know the representatives of other local agencies. They also will need to provide training and information to staff to ensure coordination with staff at other agencies, such as

- Other sectors of the local water and wastewater systems,
- Local government management,
- Local emergency responders and emergency management agencies,
- Other local utilities, particularly electrical,
- State water and wastewater primacy agencies, and
- Local public health agencies and healthcare providers.

Recent disasters also have demonstrated that local agencies need to coordinate with and accept aid from partners with new partners too, including non-governmental organizations (NGOs), such as the Red Cross, and faith-based organizations (FBO). Although relationships with NGOs and FBOs will likely be more significant for other sectors of local government, collection system managers should be familiar with them. An excellent way to become involved with other local first responder agencies is through local emergency planning committees.

3.5 Develop an Understanding of the National Incident Management System and the Incident Command System

Related to the need to develop relationships with other local agencies is the need for wastewater collection system managers to become familiar with NIMS and ICS, which is a large component of NIMS, which includes

- Command and management structure and protocols (the ICS subset of NIMS);
- Preparedness;

- Resource typing and management;
- Communications (plain English);
- Supporting technologies;
- Maintenance; and
- Training.

Both the Department of Homeland Security and Federal Emergency Management Agency (FEMA) require NIMS compliance for fund eligibility.

3.6   Be Aware of Cyber-Vulnerabilities

Cyber-system vulnerabilities should be considered in security and disaster planning. Failure of SCADA systems can result in loss of system communication with and control of wastewater pumping and lift stations and other system components. Some of the best documented intentional cyber-attacks on water and wastewater systems involve successful hacking to disable wastewater pumping and lift stations, causing SSOs.

3.7   Enter into Mutual Aid Networks and Agreements

Water and wastewater systems need to establish and develop mutual aid networks, both within and between states. Water and wastewater agencies in several states have been working to form such networks under the Water and Wastewater Agencies Response Network (WARN) and other models. The WARN establishes

- Mutual aid agreements,
- Procedures for communications and mutual aid activation, and
- Resource typing.

Mutual aid efforts in several states also have included

- Establishment of a mutual aid coordinator's position in the state emergency operations center (EOC),
- Websites to post mutual aid needs, and
- Mutual aid responders' accommodations checklists.

A significant effort is under way by WARN and other organizations and agencies to develop an interstate mutual aid framework in response to frustrations in providing aid following the Florida hurricanes in 2004 and the Gulf Coast hurricanes in 2005.

3.8 Develop Crisis Communications Plans

Wastewater collection system managers need to prepare crisis communications plans (CCP) to support interactions with the public and the media. They need to develop news releases for various scenarios within their systems. Statements that may be needed should be anticipated and practiced. Core components of a CCP process include:

- Developing the message,
- Establishing relationships with members of the media,
- Educating employees and other key stakeholders, and
- Identifying and training spokespersons and public information officers.

The U.S. Environmental Protection Agency (U.S. EPA) has been developing guidance on this task, which they refer to as "message mapping" (see http://www.epa.gov/nrmrl/news/news012006.html). Although this need is more critical need for water agencies, collection system managers also should be prepared.

3.9 Conduct Training and Exercises

Finally, wastewater collection systems should practice their disaster response plans. The U.S. Department of Homeland Security Homeland Security provides the prevailing guidance for exercise programs through its exercise evaluation program (HSEEP). The HSEEP identifies seven levels of exercises:

(1) Seminar—informal discussion led by a presenter;
(2) Workshop—formal discussion led by a facilitator;
(3) Tabletop exercise—informal discussion of a hypothetical scenario;
(4) Game—simulation of operations to depict a real-life incident;
(5) Drill—activity to test a single, specific operation or function;
(6) Functional exercise—examines coordination, command, and control between multiple-agency EOC; and
(7) Full-scale exercise—physical, interagency response to a simulated event.

One of the best ways for wastewater collection systems to practice is through relatively easy and inexpensive tabletop exercises and games, which U.S. EPA refers to as enhanced tabletop exercises. Wastewater collection system managers need to conduct tabletop exercises with increasingly broad interagency involvement, both vertically and horizontally.

A tabletop exercise involves a simulated response to a hypothetical disaster, either natural or caused by humans. Unlike other types of exercises—such as full-scale exercises and drills, which often involve the mobilization of resources to a remote location and may even involve the use of actors as victims—tabletop exercises involve working through the disaster and response activities in one or more meeting rooms. Other than personnel and presentation and supporting materials, no resources are required. Advantages of tabletop exercises over full-scale exercises include lower cost, faster planning, scheduling and organization, and less vulnerability to the weather. Tabletop exercises also are sometimes called "sandbox" or "desktop" exercises. Most tabletop exercises are conducted in six hours or less.

Tabletop exercises lend themselves particularly well to many utility systems because incidents likely to occur involve relatively abstract components and situations, such as water and wastewater contaminants. This compares to incidents that other agencies may experience such as those involving hazardous materials, in which the location and components are specific, requiring other forms of exercises.

Wastewater collection system managers should conduct exercises or to become involved exercises for incidents that may already be occurring in the area. Exercises provide the opportunity to:

- Test and practice ERPs,
- Get to know other local and emergency response agencies,
- Practice use of NIMS and ICS in interagency communications,
- Identify gaps in resources, and
- Test intraagency and interagency communications.

3.10 Stay Abreast of Developments in the Industry

Finally, wastewater collection system managers should stay abreast of incidents and developments in the industry. Much can be learned from the lessons and improvements of others. This can be accomplished through reading industry literature, attending conferences, workshops, and seminars, and networking with peers.

4.0 PHYSICAL PREPAREDNESS FOR WASTEWATER COLLECTION SYSTEMS

Physical improvements that wastewater collection system managers should make or consider for disaster preparedness include:

- Providing for emergency electrical power and/or portable pumping;

- Ensuring adequate protection of mains from waterways;
- Reinforcing aerial mains;
- Locating pumping and lift stations above flood elevations;
- Protecting vehicles and heavy equipment;
- Providing for necessary repair equipment and materials;
- Ensuring the availability of reliable communication systems; and
- Investing in a system hydraulic model to determine system capabilities and evaluate alternatives when needed.

4.1 Provide for Emergency Electrical Power

Hurricane Katrina and several other recent incidents have draw attention to the dependency of water and wastewater systems, including wastewater pumping and lift stations, on electrical power. The provision for emergency electrical power when the power supply fails is one of the most important emergency planning actions of water and wastewater systems. Wastewater collection system managers should work toward having permanent, mobile, rental, or borrowed emergency electrical generators available to keep pumping and lift stations operational. Experiences in Hurricane Katrina demonstrated that trailer-mounted generators are much more valuable for wastewater pumping and lift stations than skid-mounted units; the former can be moved from one pumping station to another.

Generators also require appropriate switchgear or generator connections and provisions for maintenance and fueling capabilities. Unlike most recommendations made so far, however, the acquisition of emergency electrical generators and associated switchgear is not inexpensive. Utilities that do not purchase generators should address this need through contracts to lease generators, provisions to borrow generators, or other arrangements. The need for emergency electrical generators is a lesson that has been demonstrated repeatedly in disasters.

Emergency electrical generators are not resources that can be obtained and then ignored. Wastewater collection systems that have invested in generators or portable pumps must provide ongoing maintenance, testing, and capability for fueling the generators.

4.2 Ensure Adequate Protection of Mains from Waterways

Perhaps the most vulnerable components of wastewater collection systems are mains located along waterways. Because of the need to optimize gravity flow in collection systems, many mains are located along streams, rivers, and other waterways. Mains in these locations are particularly vulnerable to damage from flooding and erosion. Measures that wastewater collection system managers should take to protect mains from waterways include:

- Locating the mains as far back from the waterways as possible;
- Installing manholes with rims above the flood elevation or with sealed manhole covers and vent stacks;
- Stabilizing the stream bank with rip rap stone or other measures;
- Providing regular easement maintenance to ensure access when needed; and
- Providing regular inspections of high-risk main locations, such as along waterways.

4.3 Reinforce Vulnerable Mains

The most vulnerable wastewater collection system mains are those that cross waterways, aerial mains in other low, aboveground areas, and those that are close to waterways. These mains are especially vulnerable to heavy waters, erosion, waterborne debris, and falling trees. Some wastewater collection systems have had success in reinforcing these mains by:

- Using restrained joint pipe;
- Installing the carrier with spacers ("spiders") inside an outside-welded steel casing pipe;
- Minimizing the number of concrete piers to support the pipe;
- Increasing size, depth, and footing of piers;
- Improved straps and anchors to attach the pipe to piers;
- Extending double-walled pipe farther back from stream banks; and
- Structurally reinforcing manholes.

4.4 Locate Pumping and Lift Stations above Flood Elevations

Stationary assets, such as pumping and lift stations, also should be protected as much as possible. This includes permanent improvements and temporary measures, such as:

- Raising all facilities and equipment above the potential flood elevation;
- Raising particularly vulnerable components, such as electrical equipment, above the potential flood elevation; and
- Sandbagging critical system components, like electrical components.

4.5 Protect Vehicles and Heavy Equipment

Collection system managers also must protect critical assets that are mobile. High-pressure jetting/vacuum trucks, for instance, are substantial investments for wastewater collection systems and are particularly valuable in disaster response. Wastewater collection system managers should do everything possible to put those and other critical pieces of equipment where they will be safe and available after an incident.

4.6 Provide for Necessary Repair Equipment and Materials

Wastewater collection system managers also should have on hand other resources as needed for follow-up to disasters. This may include:

- Replacement pipe,
- Replacement and portable pumps,
- Repair parts for heavy equipment,
- Sand bags and other items, and
- Gasoline and diesel fuel for vehicles and equipment.

It is also critical to protect these assets from damage or loss in a disaster.

4.7 Ensure the Availability of Reliable Communication Systems

Finally, wastewater collection system managers need to ensure the availability of functional, reliable equipment for internal communications during and after a disaster. They should maintain their investment in traditional two-way radio communications that are not dependent upon the resources of others or subject to excessive public communications traffic. Recent advances in cellular technologies have led utilities to become increasingly reliant upon those technologies. However, they have proven to be vulnerable in disasters.

Traditional radio systems may require provisions for emergency electrical power to keep them operational in a disaster. Utility agencies should consider investing in satellite phones and participation in emergency phone systems, such as the Government Emergency Telecommunications Service.

Reliable internal communications systems should not be confused with interoperable systems between agencies. Although interoperable systems are valuable, internal communication is much more critical.

5.0 DISASTER RESPONSE IN WASTEWATER COLLECTION SYSTEMS

During and after disasters, wastewater collection system managers should undertake several measures to address possible security breaches and other emergencies in the systems:

- Activate ERP and open EOCs;
- Activate crisis communications plans;
- Check on the welfare of employees;
- Conduct damage assessments;
- Activate emergency electrical power systems;
- Communicate and coordinate with other local and state response agencies using ICS;
- Activate mutual aid networks and agreements;
- Make emergency repairs; and
- Maintain records to maximize FEMA reimbursement.

5.1 Activate Emergency Response Plans and Open Emergency Operations Centers

The first thing that wastewater collection system managers should do during and after disasters and security breaches is to activate the ERPs that they have prepared and open their EOCs. This activation may include

- Contacting internal personnel and key outside agencies;
- Moving mobile and portable equipment to more secure locations, away from floodwaters and other hazards;
- Protecting critical facilities with sand bags or other measures;
- Protecting critical system information;
- Implementing succession plans or delegations of authority; and
- Establishing measures for employees to shelter-in-place at some work locations.

5.2 Activate Crisis Communications Plans

Wastewater collection system managers should use the crisis communications plans they have developed to share information with customers and stakeholders. This may include notifications of service interruptions, warnings regarding SSOs, or other key messages.

5.3 Check on the Welfare of Employees

As soon as a disaster has passed, wastewater collection system managers should strive to check on the welfare of employees and their families. Measures should be taken to provide assistance and enable the employees to participate in the agency's response to the disaster as quickly as possible.

5.4 Conduct Damage Assessments

Past disasters have demonstrated the need to make an initial assessment of the damage to a system before attempting to begin repairs. Damage to collection systems during extreme weather events often happen in areas that are not typically observed. Managers should ensure collection systems in these areas are investigated soon after the event. Damage assessments should check for several items:

- Interruptions of electrical power supplies to pumping and lift stations;
- Blockage of accessibility to facilities and areas of the collection system;
- Damage to pumping and lift stations;
- Debris over manholes;
- Damage to manholes and mains;
- Debris in manholes and mains;
- Damage to vehicles and heavy equipment; and
- Damage to supplies, materials, and other assets.

5.5 Communicate and Coordinate with Other Local and State Response Agencies

As the damage assessment is conducted, wastewater collection system managers should initiate communication and coordination with other local and state emergency response agencies, including:

- Emergency management agencies for assistance that may need to be requested in anticipation of potential FEMA reimbursement;
- State regulatory agency regarding any environmental effects that may have occurred; and
- Local electrical utility regarding restoration of power.

5.6 Activate Mutual Aid Networks and Agreements

After checking on employees, assessing the damage, and determining any resulting needs, wastewater collection system managers should request assistance through the mutual aid networks they have established. It is important to use guidelines and resource typing to request aid using established protocols and terminology. Key needs and conditions that should be addressed for mutual aid responders include:

- Housing and sanitation,
- Food and water,
- Employee work safety conditions,
- Communications systems,
- First aid and emergency medical services,
- Current inoculations required of responders,
- Safeguards for temperature ranges and weather conditions,
- Vehicles and equipment,
- Psychological support,
- Financial services, and
- Laundry services.

5.7 Make Emergency Repairs

As soon as the necessary resources can be assembled internally and through mutual aid, wastewater collection system managers should undertake necessary repairs to their system.

5.8 Maintain Records to Maximize Federal Emergency Management Agency Reimbursement

Finally, wastewater collection system managers typically will want to optimize receipt of potential reimbursement funds from FEMA. To do so, they should keep detailed records of their activities and expenses during the response.

6.0 SUGGESTED READINGS

American Society of Civil Engineers; American Waterworks Association; Water Environment Federation (2006) *Draft Guidelines for the Physical Security of Wastewater/Stormwater Utilities;* American Society of Civil Engineers: Washington, D.C.

Department of Homeland Security NRF Resource Center Home Page. http://www.fema.gov/emergency/nrf/ (accessed June 2008).

National Association of Clean Water Agencies Decontamination Wastewater e-Library. http://www.nacwa.org/index.php?option=com_content&task=view&id=47&Itemid& (accessed June 2008).

U.S. Environmental Protection Agency Water and Wastewater Security Product Guide.http://www.epa.gov/safewater/security/guide/index.html (accessed June 2008).

Index

A

Ad valorem taxes, capital funding and financing, 189
Air valves and blowoff, pressure pipelines, 118
Assessments, capital funding and financing, 189
Asset management, 3
   data, 52
   budgeting, 181, 200

B

Backfill and bedding, gravity pipelines, 112
Backfill, pressure pipelines, 117
Benchmarking, capital improvement plan, 91
Bidding, construction, 140
Blowoff and air valves, pressure pipelines, 118
Bonds
   capital funding and financing, 190
   construction financing, 150, 151
Budgeting, 173
   asset management, 181, 200
   capital funding and financing, 188
      ad valorem taxes, 189
      grants, 189
      impact fees, 189
      sources, 190
      governmental loans, 191
      revenue bonds, 190
   special assessments, 189
   user charges, 188
capital improvements, 179
   life-cycle cost analysis, 185
   underground infrastructure, 183
GASB 34, 198
general obligation bonds, 190
information management, 51
operations, 175
revenue requirements and fee setting, 191
   cash needs approach, 193
   impact fees, 197
   user charges and service fees, 195, 196
   utility approach, 194

C

Capacity, management, operations, and maintenance, 9, 100
Capital improvement plan, 59
   capacity, 61
   computer-assisted drawing, 63
   computerized maintenance management systems, 64
   condition assessment, 76
      closed-circuit television, 78
      lift stations, 81
      smoke testing, 79

Capital improvement plan,
    condition assessment *(continued)*
        surface inspection, 77
        trunkline inspection, 80
    data collection and management, 62
    flow monitoring, 69
    geographic information systems, 63
    hydraulic and hydrologic modeling, 72
    industry benchmarks, 91
    inflow and infiltration identification, 67
    prioritization, 92
    ratings, 85
        condition, 88
        performance, 88
        risk, 88
    replacement/rehabilitation goals, 91
    supervisory control and data acquisition, 63
    system analysis tools and methods, 61, 62
    validation, 90
Capital improvements budget, 179
    life-cycle cost analysis, 185
    underground infrastructure, 183
Cash flow/cost control, construction, 151
Certification, 203
Closed-circuit television
    condition assessment, 78
    inspection, construction, 153
CMOM, 9, 100
Combined sewer overflows, design, 100
Communication systems, emergency preparedness and security, 229
Community interests, 18
Community relations
    customer service, 169
    emergency preparedness, 171
    public education, 166
    public policy, 159

Computer-assisted drawing, capital improvement plan, 63
Computerized maintenance management systems
    capital improvement plan, 64
    complaints and service requests, 65
    integrating with geographic information systems, 66
    maintenance history, 64
    work orders, 66
Condition assessment
    capital improvement plan, 76
        closed-circuit television, 78
        lift stations, 81
        smoke testing, 79
        surface inspection, 77
        trunkline inspection, 80
    ratings, 88
Conflict resolution, contract documents, 148
Constructability review, contract documents, 146
Construction contract documents, 146
    all location at risk, 147
    conflict resolution, 148
    constructability review, 146
    dispute resolution, 148
    escrow documents, 148
    operations and maintenance review, 146
    value engineering, 146
Construction
    agency management, 145
    alliances and partnering, 155
    bonds, 151
    cash flow/cost control, 151
    competitive bidding, 140
    contracting, 137
    environmental requirements, 141
    funding mechanisms, 150
        grants and loans, 150
        municipal bonds, 150

inspection/administration, 153
insurance, 151
interagency permit requirements, 141
laws and regulations, 139
maintenance project delivery, 145
   force account construction, 145
   term contracts, 145
payment methods, 148
   cost-plus basis, 149
   lump-sum basis, 149
   unit-price basis, 149
project delivery, 142
   at-risk construction management, 144
   design–bid–build, 142
   design–build, 143
   multiple prime contractors, 142
project management, 138
quality assurance/quality control, 152
safety, 154
utility damage control (one-call systems), 154
warranty/guarantee administration, 154
Contract documents
   all location at risk, 147
   conflict resolution, 148
   constructability review, 146
   construction, 146
   dispute resolution, 148
   escrow documents, 148
   operations and maintenance review, 146
   value engineering, 147
Contracting, construction, 137
Conveyance rates
   design, 101
   risk-based design, 102
   wet-weather facilities, 102
Corrosion control and odor, design, 119
Corrosion protection, design, 122

Cost control/cash flow, construction, 151
Covers, manholes, 128
Crisis communication, 225, 230
Crossings and tunnels, design, 133
Customer complaints and service requests, computerized maintenance management systems, 65
Customer expectations/complaints, 17
Customer service and community relations, 169
Cyber vulnerabilities, 224

D

Data collection and management, capital improvement plan, 62
Definitions, 6
Design, 97
   approvals, 102
   capacity, management, operations, and maintenance, 100
   consultants, 105
   corrosion protection, 122
   crossings and tunnels, 133
   environmental issues and compliance, 103
   gravity pipelines, 110
   guidelines, 103
   hydraulic analysis using modeling, 108
   infiltration and inflow, 125
   inverted siphons, 132
   manhole and pipeline testing, 130
   manholes, 125
   National Pollutant Elimination System Permits, 100
   odor and corrosion control, 119
   permitting, 103

Design *(continued)*
　pipelines and joint materials, 110
　pressure pipelines, 115
　process checklist, 103
　project management, 104
　pumping stations, 109
　regulatory and environmental
　　　requirements, 100
　rehabilitation, 123
　risk-based, 102
　routing considerations, 105
　　aboveground and background
　　　interference, 105
　　geotechnical, 106
　　permanent access, 107
　　potholing, 106
　　public effects, 107
　　rights-of-way and easements, 108
　　surveys, 106
　　traffic control, 106
　sanitary sewer overflows and
　　　combined sewer overflows, 100
　sanitary sewers above grade, 133
　service connections and
　　　disconnections, 134
　specification writing, 135
　system conveyance rates, 101
　system, 97
　technologies, 124
　value engineering, 135
　wet-weather facilities, 102
Design–bid–build, construction, 142
Design–build, construction, 143
Disaster response, 230
Dispute resolution, contract
　　documents, 148

E

Easements and rights-of-way, design, 108
Electrical power, emergency preparedness
　　and security, 227
Emergency preparedness and security,
　　171, 217
　communication systems, 229
　crisis communication plans, 225
　cyber vulnerabilities, 224
　damage assessment, 231
　disaster response, 230
　emergency electrical power, 227
　emergency response plans and
　　　operations centers, 230
　equipment repair, 229
　FEMA, 232
　Incident, Command System, 223
　incidents caused by humans, 219
　National Incident Management
　　　System, 223
　natural disasters, 219
　protection of mains from
　　　waterways, 227
　pumping and lift stations, 228
　technological failures, 220
　training, 225
Emergency response, 230
Environmental requirements,
　　construction, 141
Equipment
　health and safety, 215
　needs, operations and
　　　maintenance, 15
　operations and maintenance, 44
　repair, emergency preparedness and
　　　security, 229
Escrow, contract documents, 148

F

Federal Emergency Management
　　Agency, 232
Fee setting and revenue requirements,
　　budgeting, 191
　cash needs approach, 193
　impact fees, 197
　user charges and service fees, 195, 196
　utility approach, 194
Financial considerations, construction, 148
Financial planning and budgeting, 173

Financing and funding
　ad valorem taxes, 189
　capital, 188
　grants, 189
　impact fees, 189
　sources, 190
　　general obligation bonds, 190
　　governmental loans, 191
　　revenue bonds, 190
　　special assessments, 189
　　user charges, 188
First-aid, 213
Flow monitoring, capital improvement plan, 69
Funding mechanisms, construction, 150

G

Geographic information systems
　capital improvement plan, 63
　integrating with computerized maintenance management systems, 66
Glossary, 6
Government Accounting Standards Board Statement Number 34, 198
Grades, gravity pipelines, 114
Grants
　capital funding and financing, 189
　construction financing, 150
Gravity pipelines, design, 110
Guarantee/warranty administration, construction, 154

H

Health and safety
　employee orientation and training, 210
　equipment, 215
　first-aid, 213
　policy, 205
　program, 204, 208, 214
　rules and regulations, 204, 208
　workplace inspections, 212
Hydraulic and hydrologic modeling, 72
　calibration/verification, 75
　flow development, 74
　steady-state simulation versus dynamic simulation, 73

I

Impact fees
　capital funding and financing, 189
　revenue requirements and fee setting, budgeting, 197
Incident Command System, 223
Infiltration and inflow, design, 125
　allowance, gravity pipelines, 114
　identification, capital improvement plan, 67
Information management, 47
　budgeting, 51
　guidelines, 57
　operations and maintenance, 58
　preventive maintenance and inspection calendars, 51
　regulatory responsibilities, 49
　searches, 50
　system attributes, 54
　types of collection system information, 53
　work order and repair schedules, 52
Inspections
　administration, construction, 153
　calendars, 51
　closed-circuit television, 78
　health and safety, 212
　smoke testing, 79
　surface, 77
　trunkline, 80

Insurance, construction, 151
Inverted siphons, design, 132

J

Joint materials, design, 110

L

Laws and regulations, construction, 139
Life-cycle cost analysis, capital improvements budgeting, 185
Lift and pumping stations, emergency preparedness and security, 228
Lift stations, condition assessment, 81
Loans
  capital funding and financing, 191
  construction financing, 150

M

Maintenance, 23, 27
  history, computerized maintenance management systems, 64
  project delivery, 145
    force account construction, 145
    term contracts, 145
  pumping stations, 34, 36, 37
  pumping station staffing, 34, 40, 42, 43
Manhole, testing, 130
Manholes, design, 125
Mapping, 27
Materials
  gravity pipelines, 111
  pressure pipelines, 116
Modeling
  hydraulic analysis, 108
  hydraulic and hydrologic, 72
    calibration/verification, 75
    flow development, 74
    steady-state simulation versus dynamic simulation, 73
Municipal bonds, construction funding, 150

N

National Incident Management System, 223
National Pollutant Elimination System Permits, design, 100
Natural disasters, 219
NIMS, 223

O

Occupational Safety and Health Administration, 203
Odor and corrosion control, design, 119
Operations and maintenance, 13
  agency resources, 15
  community interests, 18
  customer expectations/complaints, 17
  equipment needs, 15
  facilities and equipment, 44
  information system, 58
  maintenance, 23, 27
  mapping, 27
  performance history, 15
  problems and overflow response, 31
  program components, 19
  program objectives, 14
  pumping station staffing, 34, 40, 42, 43
  pumping stations, 34, 40, 42, 43
  regulatory resources, 19
  review, contract documents, 146
  staffing, 15, 34, 43
  system asset characteristics, 14
  technology, 29

Operations, budget, 175
Operations, pumping station staffing, 34, 40, 42, 43
Overflow response, 31

P

Performance, ratings, 88
Permit requirements, construction, 141
Permitting, design, 103
Pipelines, design, 110
Policy, heath and safety, 205
Pressure pipelines, design, 115
Preventive maintenance, calendars, 51
Project delivery
   maintenance, 145
   methods, construction, 142
Project management, construction, 138
Public education and community relations, 166
Public policy and community relations, 159
Pumping and lift stations, emergency preparedness and security, 228
Pumping stations
   design, 109
   maintenance, 34, 36, 37
   operations and maintenance, 34
   staffing, 34, 40, 42, 43
     maintenance, 43
     operations, 42

Q

Quality assurance/quality control, construction, 152

R

Regulations and laws, construction, 139
Regulations, health and safety, 204
Regulatory and environmental requirements, design, 100
Regulatory resources, 19
Regulatory responsibilities, information management, 49
Rehabilitation
   design, 123
   manhole and pipeline testing, 131
   manholes, 127
Rehabilitation/replacement goals, capital improvement plan, 91
Repair schedules, 52
Replacement/rehabilitation goals, capital improvement plan, 91
Restrained joints, pressure pipelines, 117
Revenue requirements and fee setting, budgeting, 191
   cash needs approach, 193
   impact fees, 197
   user charges and service fees, 195, 196
   utility approach, 194
Rights-of-way and easements, design, 108
Risk, ratings, 88
Routing considerations, design, 105

S

Safety and health, 203
   committee, 207
   construction, 154
   employee orientation and training, 210
   equipment, 215
   first-aid, 213
   job hazard analysis, 208
   policy, 205
   program, 204, 208, 214
   rules and regulations, 204, 208
   workplace inspections, 212
Sanitary sewer overflows, design, 100
Sanitary sewers above grade, design, 133

Security and emergency preparedness, 217
　communication systems, 229
　crisis communication plans, 225
　cyber vulnerabilities, 224
　damage assessment, 231
　disaster response, 230
　emergency electrical power, 227
　emergency response plans and operations centers, 230
　equipment repair, 229
　FEMA, 232
　Incident Command System, 223
　incidents caused by humans, 219
　natural disasters, 219
　protection of mains from waterways, 227
　pumping and lift stations, 228
　technological failures, 220
　training, 225
　National Incident Management System, 223
Service connections and disconnections, design, 134
Sizing considerations
　gravity pipelines, 110
　pressure pipelines, 116
Smoke testing, condition assessment, 79
Specification writing, design, 135
Staffing
　operations and maintenance, 15
　pumping stations, 34, 40, 42, 43
Standard operating procedures, 203, 206, 210
Sulfide control, design, 120
Supervisory control and data acquisition, capital improvement plan, 63

T

Technologies for design, 124
Technology, 29
Term contracts, maintenance project delivery, 145
Terminology, 6
Testing, manholes and pipelines, 130
Training, 203
　emergency preparedness and security, 225
　health and safety, 210
Trunkline inspection, condition assessment, 80
Tunnels and crossings, design, 133

U

User charge and service fee types, 196
　capital funding and financing, 188
　revenue requirements and fee setting, budgeting, 195, 196

V

Value engineering
　contract documents, 146
　design, 135

W

Warranty/guarantee administration, construction, 154
Water and Wastewater Agencies Response Network, 224
Wet-weather facilities, design, 102
Work order and repair schedules, 52
Work orders, computerized maintenance management systems, 66